Atlas of
Ultrastructure
Ultrastructural Features
In Pathology

ATLAS OF ULTRASTRUCTURE
Ultrastructural Features In Pathology

Edited by C. Howard Tseng, M.D., Ph.D.

Chief, Laboratory Service
Veterans Administration West Side Medical Center
Chicago, Illinois

Associate Professor of Pathology
University of Illinois, Abraham Lincoln School of Medicine
Chicago, Illinois

Present affiliation:
Chief, Laboratory Service
Veterans Administration Medical Center
Fargo, North Dakota

Professor of Pathology
University of North Dakota
Grand Forks, North Dakota

APPLETON-CENTURY-CROFTS/New York

80 81 82 83 84 / 10 9 8 7 6 5 4 3 2 1

Prentice-Hall International, Inc., London
Prentice-Hall of Australia, Pty. Ltd., Sydney
Prentice-Hall of India Private Limited, New Delhi
Prentice-Hall of Japan, Inc., Tokyo
Prentice-Hall of Southeast Asia (Pte.) Ltd., Singapore
Whitehall Books Ltd., Wellington, New Zealand

Library of Congress Cataloging in Publication Data
Main entry under title:

Atlas of ultrastructure.

   1. Diagnosis, Electron microscopic—Atlases. 2. Ultrastructure (Biology)—Atlases.
I. Tseng, C. Howard, 1938-
RB43.5.A86   616.07′5   79-20106
ISBN 0-8385-0462-0

Text design: Holly Reid
Cover design: Susan Rich

PRINTED IN THE UNITED STATES OF AMERICA

CONTRIBUTORS

Nobuhisa Baba, M.D., Ph.D.

Director, Division of Pathologic Anatomy, University Hospital; Professor of Pathology, Ohio State University College of Medicine, Columbus, Ohio

A. Joseph Brough, M.D.

Chief, Department of Laboratory Medicine, Children's Hospital of Michigan; Associate Professor of Pathology, Wayne State University School of Medicine, Detroit, Michigan.

Robert J. Buschmann, Ph.D.

Supervisor, Electron Microscopy Laboratory, Veterans Administration West Side Medical Center; Assistant Professor of Pathology, University of Illinois, Abraham Lincoln School of Medicine, Chicago, Illinois

Immacula Cantave, M.D.

Staff Pathologist, Laboratory Service, Veterans Administration West Side Medical Center; Assistant Professor of Pathology, University of Illinois, Abraham Lincoln School of Medicine, Chicago, Illinois

Chung-Ho Chang, M.D.

Director, Division of Electron Microscopy, Children's Hospital of Michigan; Assistant Professor of Pathology, Wayne State University School of Medicine, Detroit, Michigan

Yeun Sook Choi, M.D.

Staff Pathologist, Laboratory Service, Veterans Administration West Side Medical Center; Assistant Professor of Pathology, University of Illinois, Abraham Lincoln School of Medicine, Chicago, Illinois

Contributors

Luna Ghosh, M.D.

Assistant Professor of Pathology and Surgery, University of Illinois, Abraham Lincoln School of Medicine, Chicago, Illinois

E. P. Grogg, M.D.

Director of Laboratories, Carle Clinic and Carle Foundation Hospital; Clinical Assistant Professor of Pathology, University of Illinois School of Basic Medical Sciences and Clinical Medicine, Urbana, Illinois

Teng-Liang Huang, M.D.

Chief, Orthopedic Section, Veterans Administration West Side Medical Center, Assistant Professor of Orthopedic Surgery, University of Illinois, Abraham Lincoln School of Medicine, Chicago, Illinois

Jose R. Manaligod, M.D., Ph.D.

Associate Professor of Pathology, University of Illinois, Abraham Lincoln School of Medicine, Chicago, Illinois

Juan Mir, M.D.

Assistant Chief, Laboratory Service, Veterans Administration West Side Medical Center, Assistant Professor of Pathology, University of Illinois, Abraham Lincoln School of Medicine, Chicago, Illinois

Salve G. Ronan, M.D.

Assistant Professor of Pathology and Dermatology, University of Illinois, Abraham Lincoln School of Medicine, Chicago, Illinois

Sudarshan Sahgal, M.D.

Staff Pathologist, Laboratory Service, Veterans Administration West Side Medical Center; Assistant Professor of Pathology, University of Illinois, Abraham Lincoln School of Medicine, Chicago, Illinois

C. Howard Tseng, M.D., Ph.D.

Chief, Laboratory Service, Veterans Administration West Side Medical Center; Associate Professor of Pathology, University of Illinois, Abraham Lincoln School of Medicine, Chicago, Illinois

CONTENTS

Contents

Contents

Contents

Contents

19   BONE AND JOINT

20   RETICULOENDOTHELIAL SYSTEM

Contents

PREFACE

This concise atlas contains precise illustrations of diagnostic features of particular diseases as well as the manifestations of common or specific pathologic processes.

Many electron microscopy atlases illustrate either normal human and animal ultrastructure or give extensive treatises on specific pathology subjects. Our atlas, however, is designed to be a comprehensive, easily read and assimilated atlas of routine electron micrographs that will provide a general introduction and review so needed by novices and by others interested in the field.

This atlas includes 200 illustrations from biopsy and surgical specimens from several large medical centers having routine electron microscopy services for the last decade. In addition, most of the electron microscopic diagnostic features of the diseases are included in this atlas.

The atlas is intended to serve as a desk reference for electron microscopists and as a comprehensive review for physicians preparing for various speciality board examinations. The atlas will also provide medical students and pathology residents with a basic understanding and the most current knowledge of human organ ultrastructure in abnormal conditions.

We acknowledge the insight and enterprise of Dr. Marjorie J. Williams who as Director of the Pathology Service of the Veterans Administration initiated and has continuously promoted diagnostic electron microscopy in the Veterans Administration. We express our gratitude to Dr. Roy J. Korn and to Dr. Samuel T. Nerenberg who have encouraged and supported this endeavor. We thank Drs. Charles F. Wooley, Donald J. Unverferth, and John R. Svirbely for their contributions. We are indebted to Jeffrey L. Harrison, Jerome M. Johnson, Nearl R. Mailman, and Fred Uchwat for their technical contributions.

1 Heart and Blood Vessel

Cardiomyopathy

FIGURE 1. This electron micrograph shows clumping of chromatin and marked invagination of nuclear membrane. (13,000×)

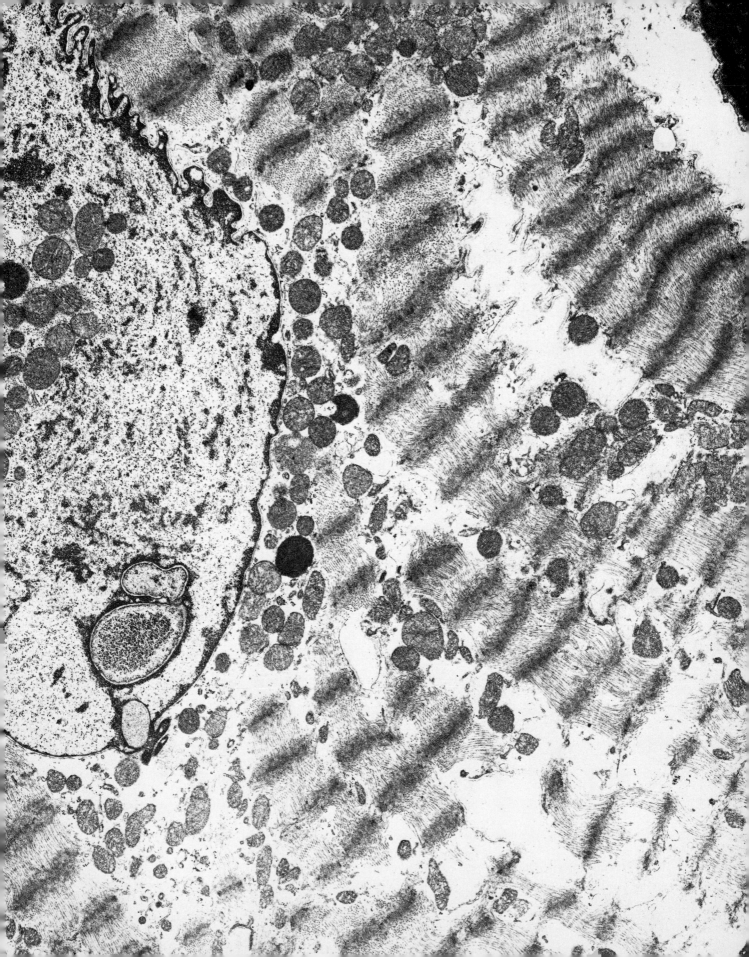

Cardiomyopathy

FIGURE 2. This micrograph demonstrates rare mitochondria and dilated cisternae of the sarcoplasmic reticulum. (17,700×)

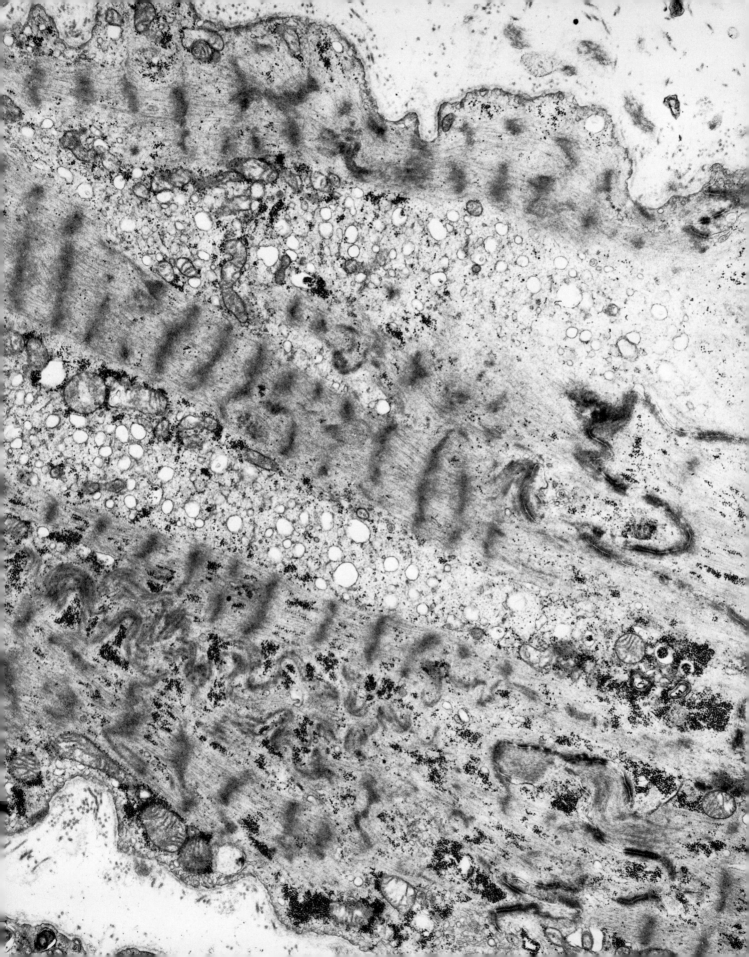

Cardiomyopathy

FIGURE 3. Shown here are increased lipochrome pigments. (36,000×)

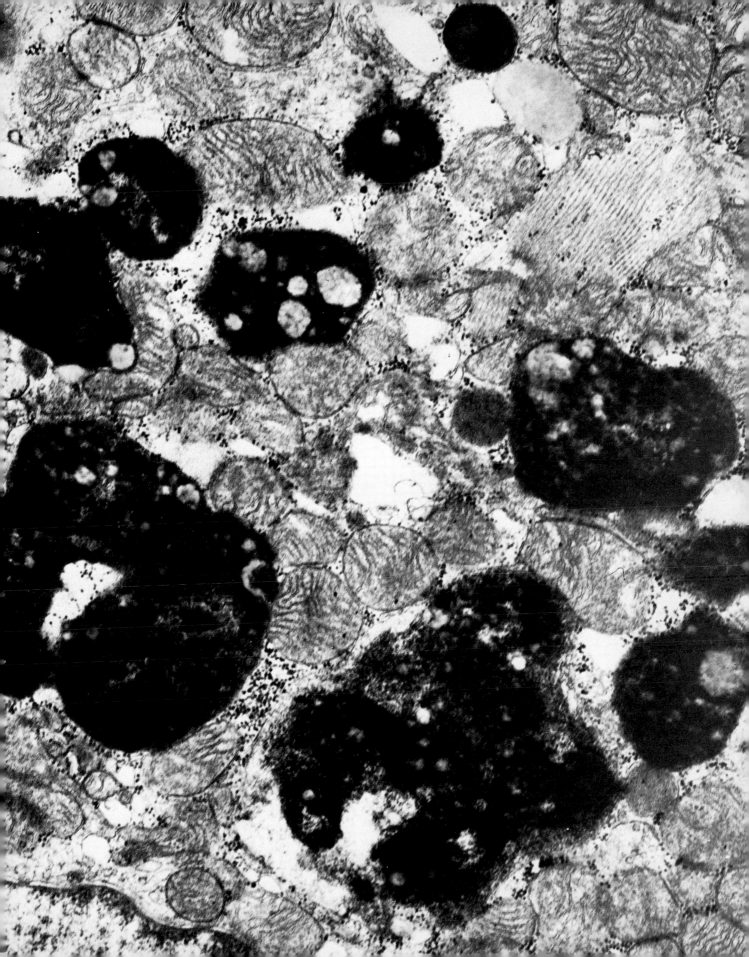

Cardiomyopathy

FIGURE 4. Abnormally increased sarcoplasmic reticulum and irregularly oriented intercalated discs are shown in this micrograph. Increased lipochrome pigments are also seen. (26,000×)

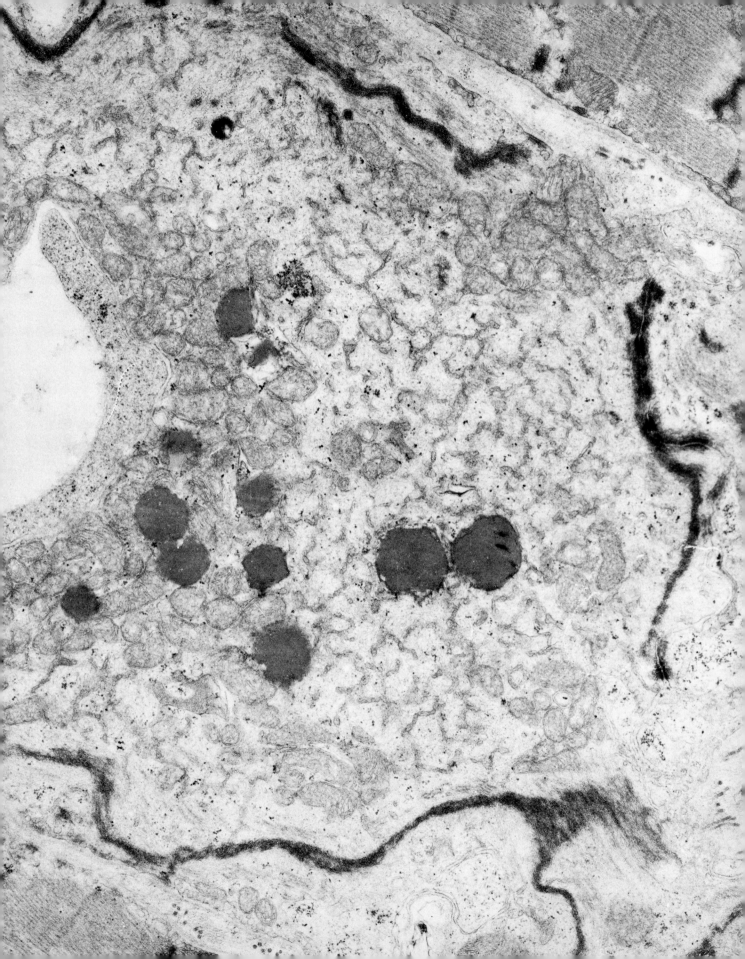

Cardiomyopathy

FIGURE 5. This electron micrograph demonstrates markedly increased lipochrome pigments. There are large areas filled with disrupted myofibrils. (14,400×)

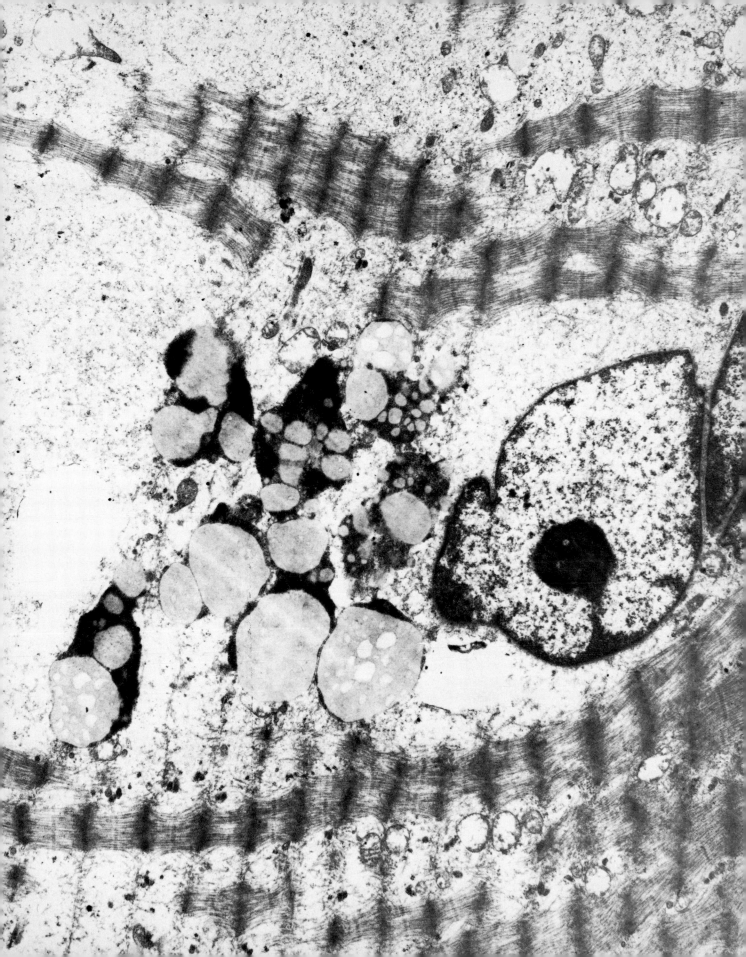

Cardiomyopathy

FIGURE 6. Revealed in this electron micrograph are disrupted myofibrils and irregular myofibrillary directions. There is increased Z-line material. (67,200×)

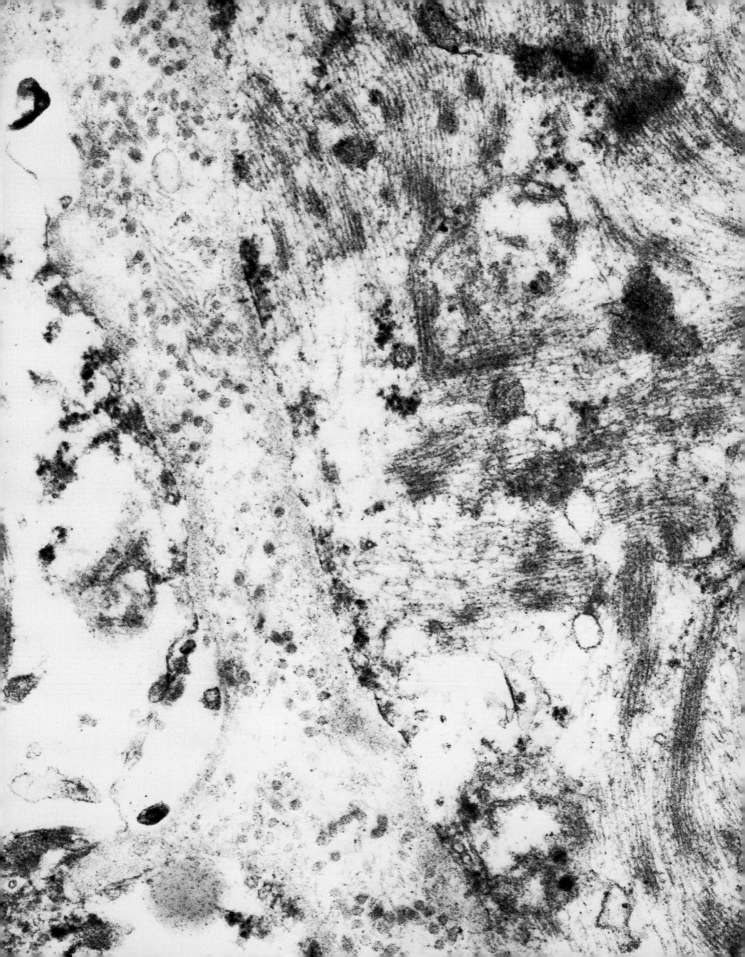

Congestive Cardiomyopathy

FIGURE 7. Pictured here are cardiac muscle fibers from a 7-year-old female with diffuse cardiac enlargement and progressive cardiac failure. Note that the myofibrils are widely separated by excessive aggregates of mitochondria. The lamellar structure of myofibrils remains well preserved. In addition to a loss of dense granules in most of the mitochondria, there is some residual glycogen in the intermyofibrillar spaces. The etiology of congestive cardiomyopathy is not clear, but its clinical onset is frequently associated with viral illness. (21,500×)

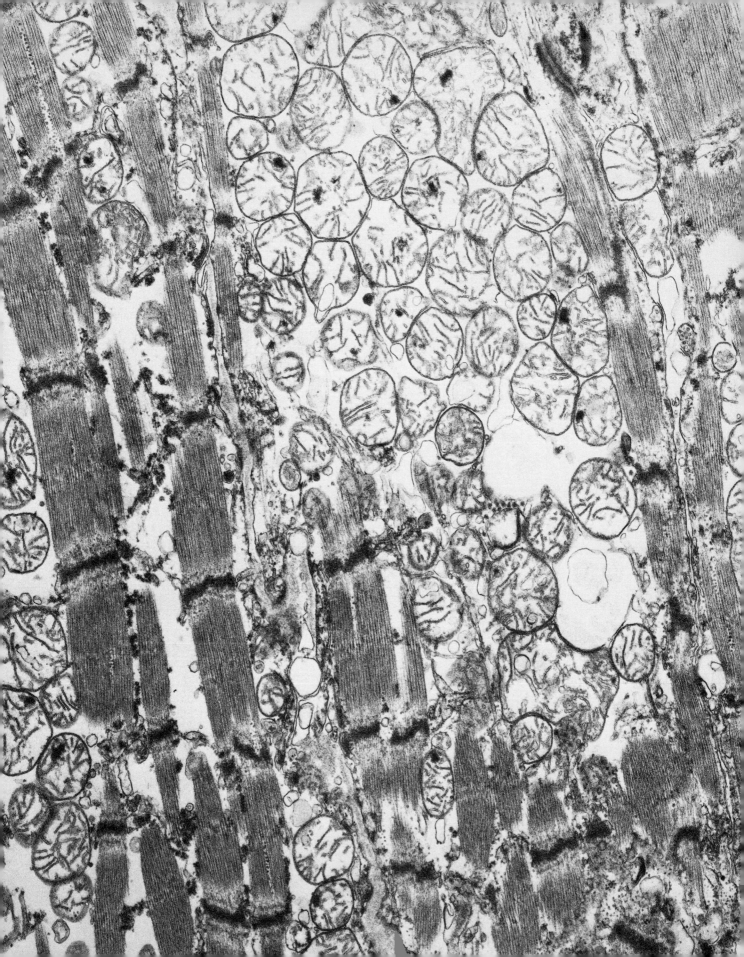

Cardiac Muscle in Glycogenosis, Type 11

FIGURE 8.   This electron micrograph details a portion of cardiac muscle fiber from the preceding case. Note the multiple cytosomes with an outer limiting membrane (arrow) containing glycogen particles in this extremely disorganized cell. Some of the glycogen particles have been extracted and one of the cytosomes appears empty. Residual myofibrils are seen in the periphery of the cell. (47,500×)

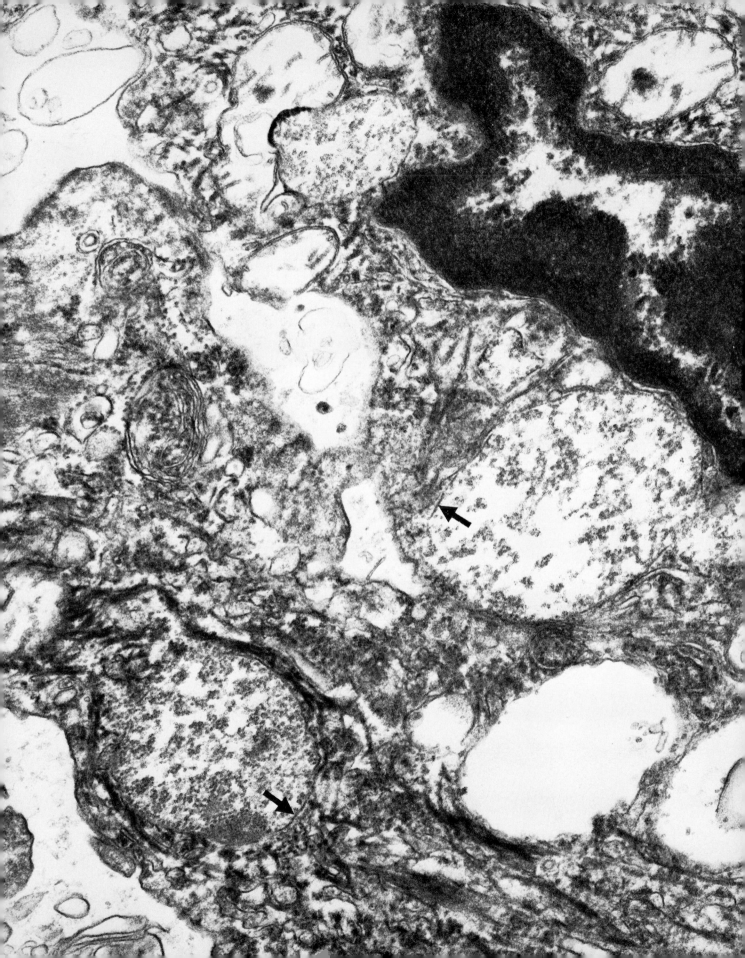

Rhabdomyoma of the Heart Associated With Tuberous Sclerosis

FIGURE 9. Portions of rhabdomyoma cells from a full-term newborn who presented with a cardiac murmur at birth. Note that the cardiac muscles contain abundant glycogen with peripheral displacement of the nucleus (N) and myofibrils (mf). The baby was found to have multiple small subependymal tubercles characteristic for tuberous sclerosis at postmortem examination. Rhabdomyoma is the most common congenital cardiac tumor and approximately half of the cases are found to be associated with tuberous sclerosis. (18,500×)

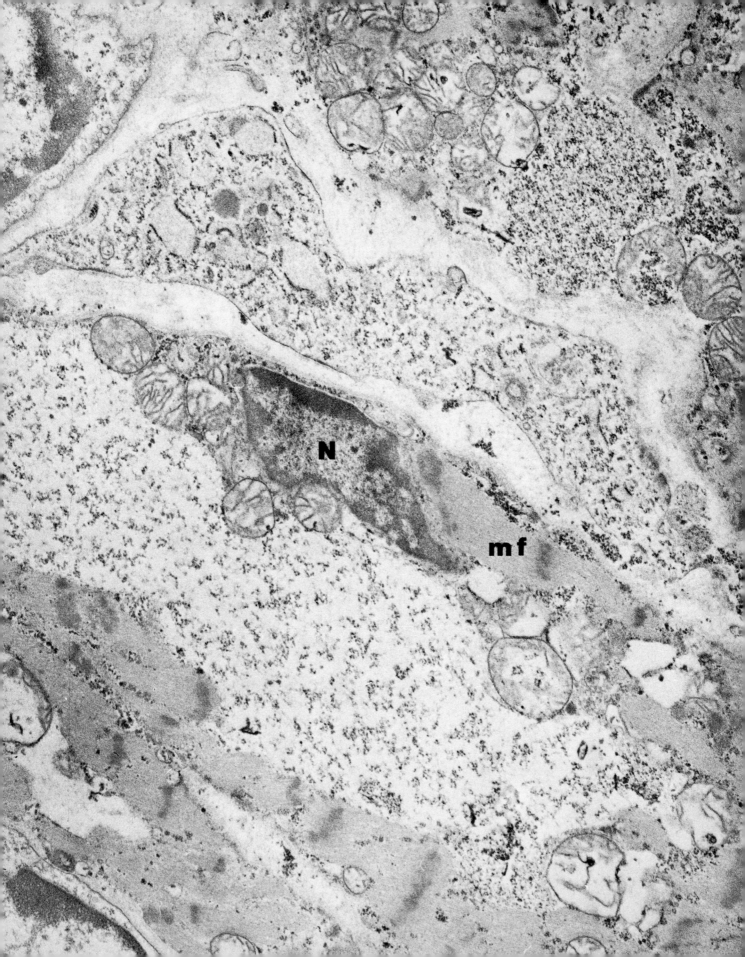

Rod-Shaped Tubulated Bodies (Weibel-Palade Bodies) In a Capillary Endothelium

FIGURE 10. Endothelial cells have scattered single-membrane-bound rod-like bodies, called Weibel-Palade bodies. They contain microtubules in an electron-dense matrix which generally have a straight parallel course along the long axis of the organelle (not shown clearly at the magnification of this picture). Occasionally pinocytotic vesicles are also characteristic of endothelial cells. These features are useful in the diagnosis of angiosarcoma —the neoplasm of endothelial cells. (15,400×)

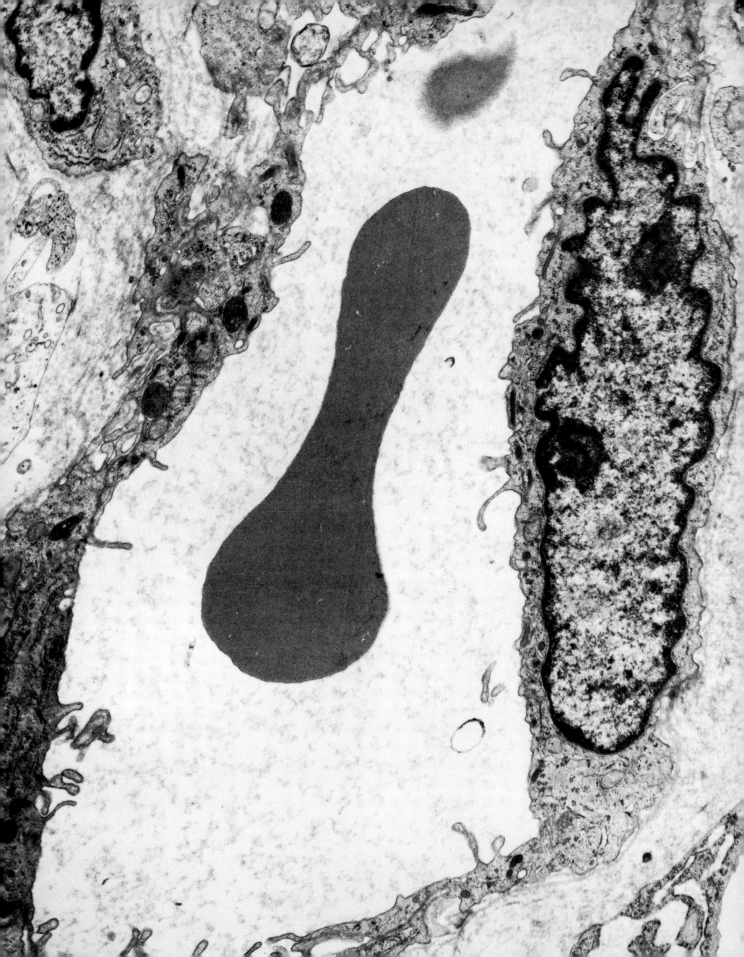

Glomus Tumor of the Finger

FIGURE 11. Panoramic view of a glomus tumor showing a concentric layering of the tumor cells around a fenestrated capillary. The innermost tumor cells are flattened, elongated, closely placed; while the peripheral cells are plump, oval to polyhedral and separated by a broader extracellular space containing mainly collagen fibrils. A well-developed basal lamina of uneven thickness invests each individual cell. Numerous pinocytotic vesicles are observed along the cell membrane. The number of cytoplasmic organelles which include mitochondria, RER, free ribosomes, lysosomes is inversely proportional to the filament content of the cell. (9800×)

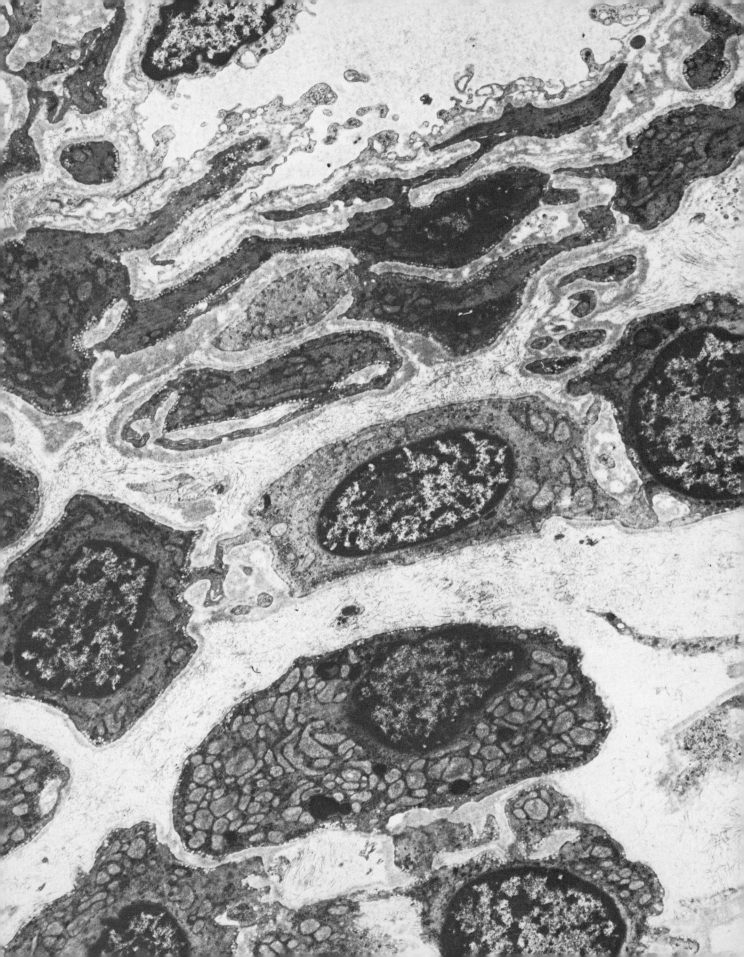

2 Lung

Type II (or Granular) Pneumocyte

FIGURE 12. Electron micrograph of portions of two alveoli from a child with viral pneumonitis. Note a proliferation of type II pneumocytes. They are united to the adjacent squamous type cells (type I cell) by tight junctions. The type II pneumocytes are cuboidal with a central nucleus. There are a few short cytoplasmic processes on the luminal surface. The characteristic features of these cells are the presence of lamellar bodies and richness in secretory granules. Note that abundant Golgi saccules are seen near the nuclei. The alveoli are separated from capillaries by narrow fibrous septae. (15,500×)

24

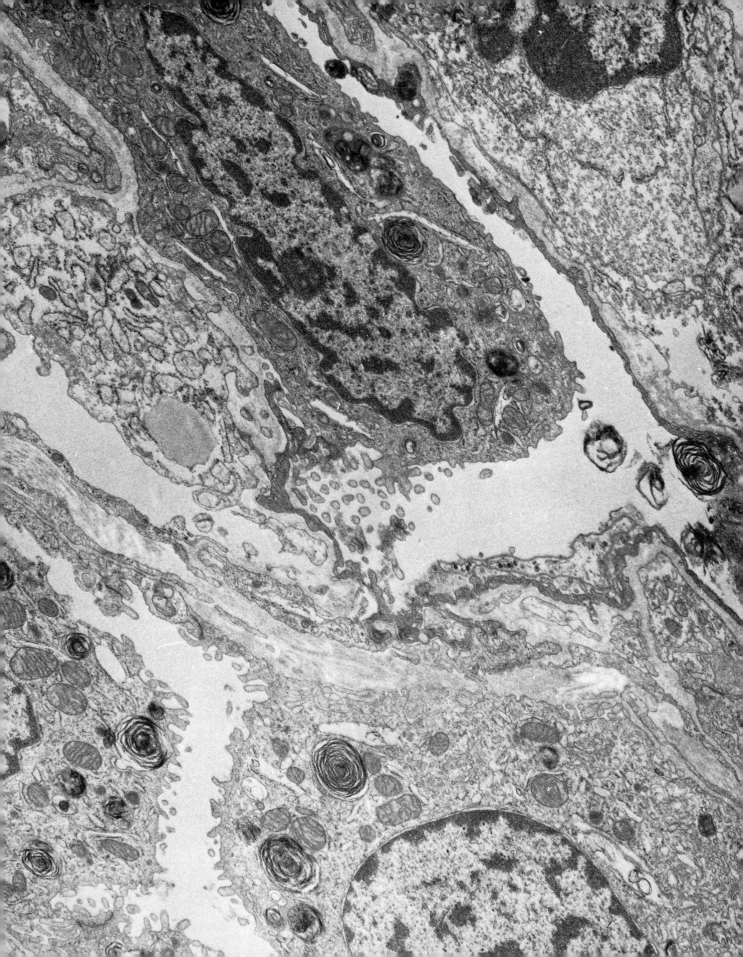

Eosinophilic Granuloma of the Lung
(Primary Pulmonary Histiocytosis X)

FIGURE 13. The cellular infiltrate in the lungs in eosinophilic granuloma is composed mainly of histiocytes, lymphocytes, eosinophils, and plasma cells. The histiocytes contain the so-called Langerhans granules which are rods with terminal vesicles resembling "tennis rackets." (19,800×)

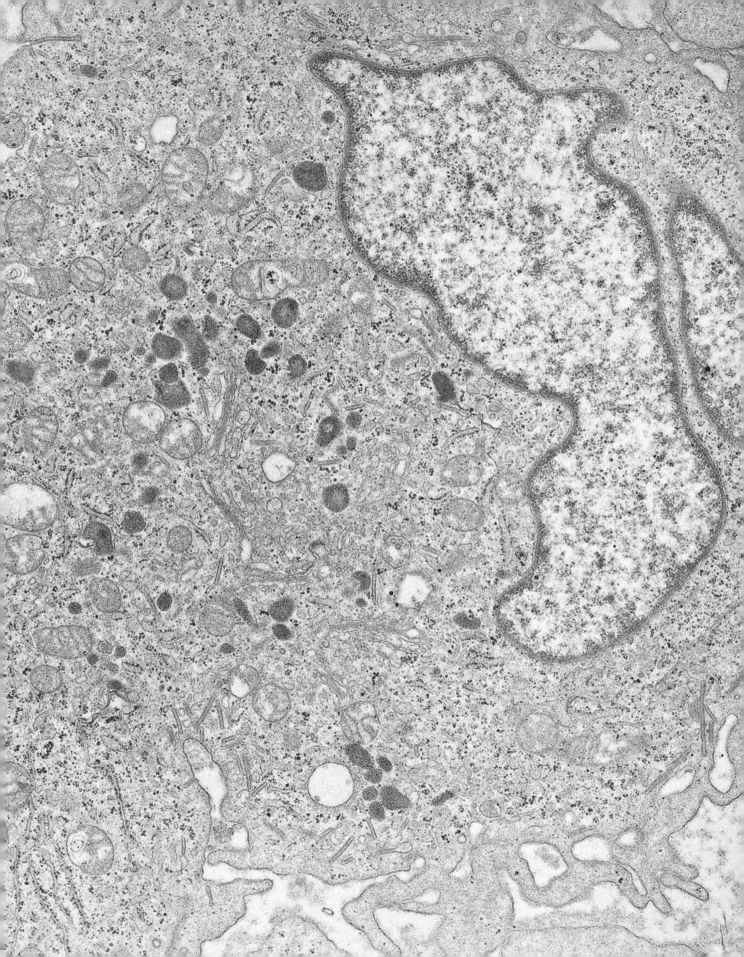

Langerhans Cell Granules in Eosinophilic Granuloma

FIGURE 14. Most of the Langerhans cell granules appear as rods, but some have a "tennis-racket" appearance. The rod has a central lamella surrounded by a trilaminar membrane. These granules are morphologically indistinguishable from those seen in the Langerhans cells in the normal epidermis. Other conditions that show the same type cells are Letterer-Siwe disease and Hand-Schüller-Christian disease. (61,600×)

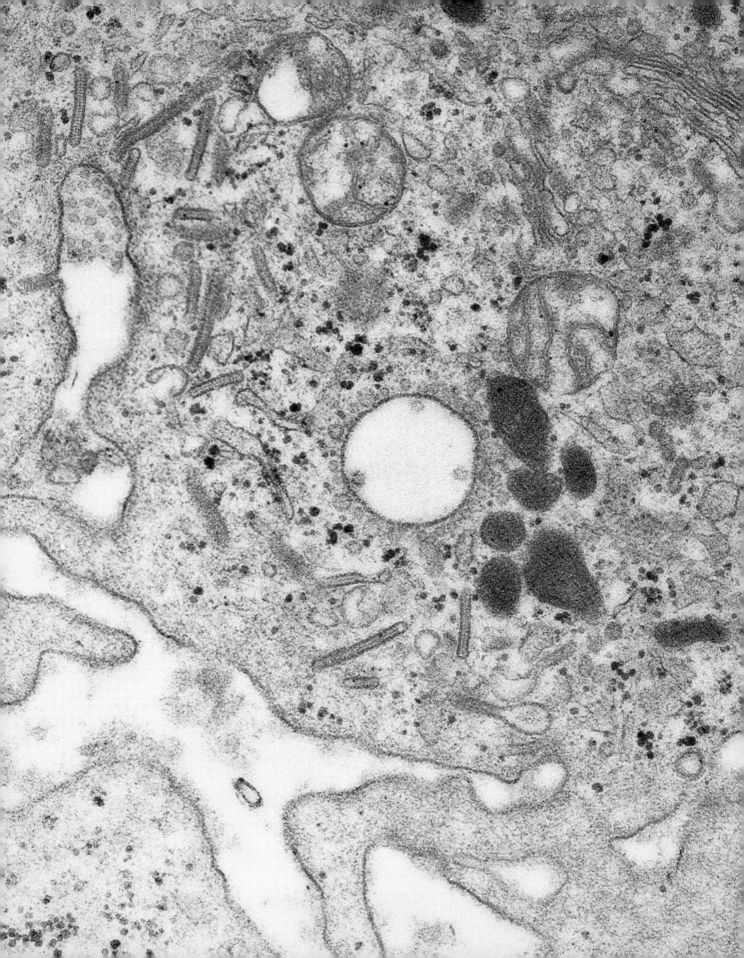

Bronchogenic Carcinoma

FIGURE 15. Secretory product of moderate electron density is seen in an irregular microacinar lumen into which delicate microvilli project. The lining neoplastic cells are pleomorphic and haphazardly oriented without any polarity. The various organelles dispersed in the cytoplasm fail to show any distinguishing features. Junctional complexes are frequently seen between adjacent cells. The nuclei contain euchromatin and nucleoli. Clinically the middle lobe of the lung was resected for a small $2.0 \times 1.0 \times 1.0$ cm peripheral lesion. By light microscopy a poorly differentiated squamous cell carcinoma was entertained, the ultrastructural study shows the glandular formations that are indicative of an adenocarcinoma. ($8500\times$)

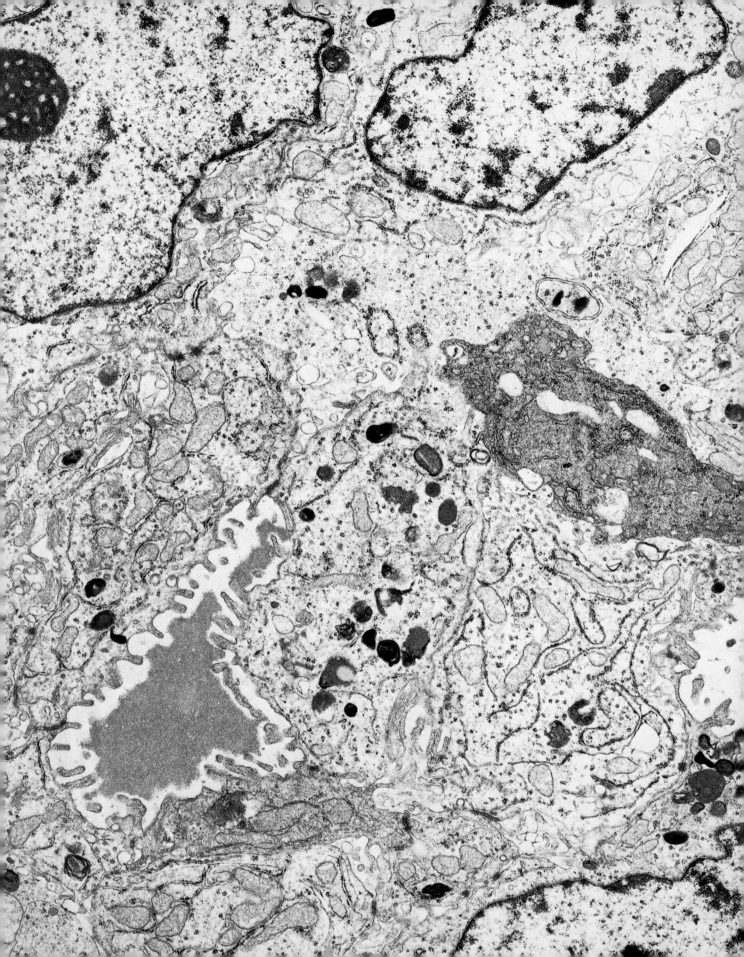

Adenocarcinoma of the Lung

FIGURE 16. The tumor tissue is composed of polygonal cells frequently connected by desmosomes. The cells contain scattered or aggregated electron dense granules. These granules are usually 1 μ or less in diameter, but irregular in shape and size. They show distinct internal myelin figures of fingerprint patterns. This internal structure cannot be appreciated at this low magnification. A higher magnification is given in subsequent figure. Kimula (Am J Surg Path 2:243, 1978) observed that these cytological characteristics resembled those of the normal nonciliated bronchiolar cells. (4900×)

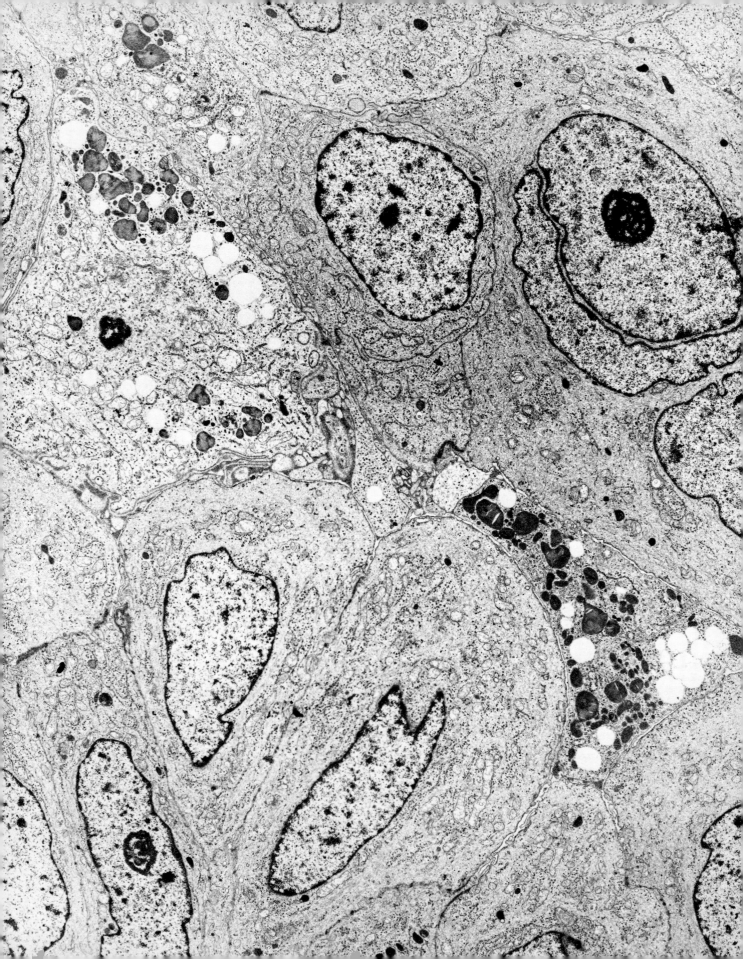

Adenocarcinoma of the Lung

FIGURE 17.   A fairly circular glandular lumen is delineated by neoplastic cells that are connected by an apparent zonula occludens. Microvilli project from the apical cytoplasm into the lumen. Aggregates of small granules of various electron densities not only occupy the apical and lateral portions of the cytoplasm but also are in close relation with a well-developed Golgi complex, and probably represent secretory granules or lysosomes. The endoplasmic reticulum appears as irregular lamellae distributed throughout the cytoplasm and bears sparse patches of ribosomes. Ribosome particles are conspicuous. A few mitochondria can be seen. (23,200×)

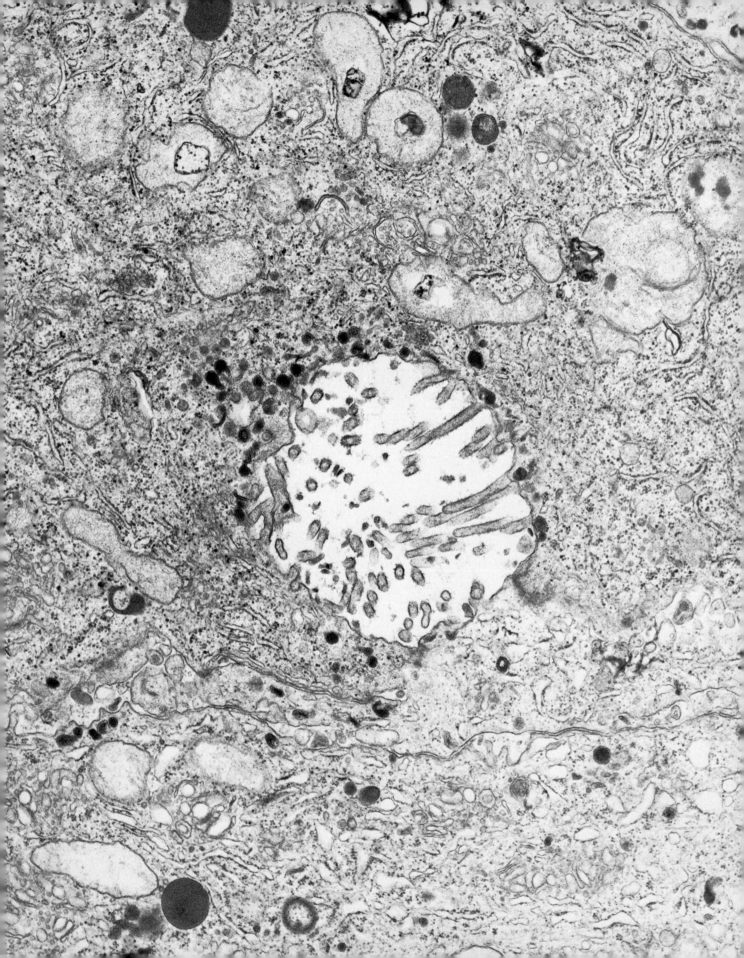

Adenocarcinoma of the Lung

FIGURE 18. The higher magnification of the previous figure shows details of the granules. They are membrane-bound, electron dense and reveal a regular array of straight membrane plates. (34,600×)

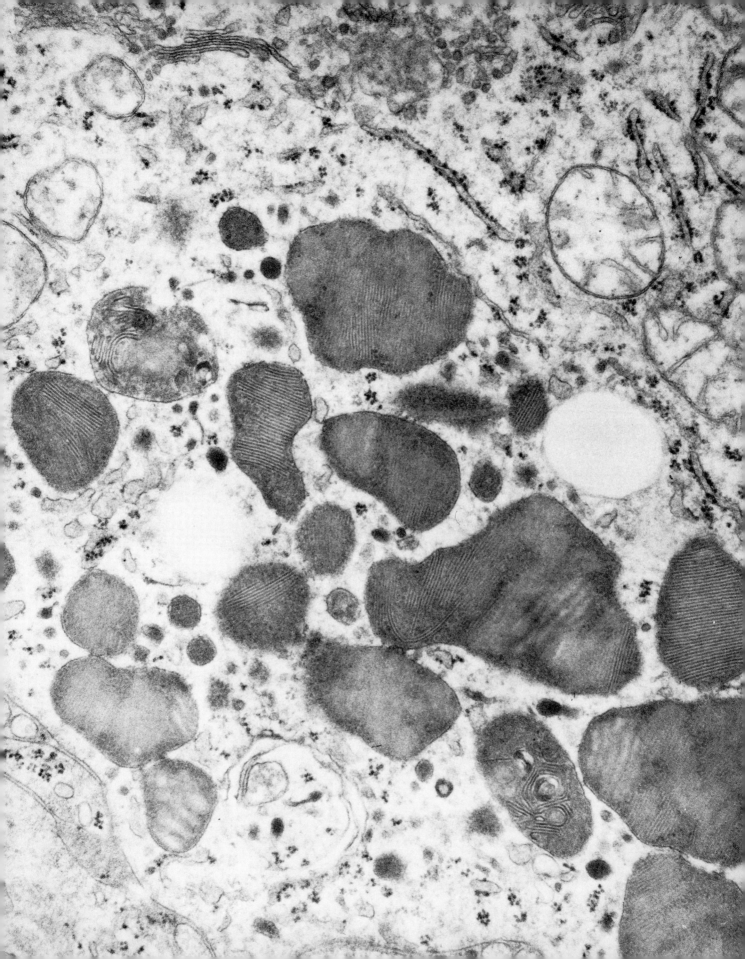

Squamous Cell Carcinoma of the Lung

FIGURE 19.   The scant cytoplasm surrounding the nucleus contains bundles of filaments several of which are anchored to the desmosomes. Other than these relatively specific characteristics of squamous cells, the cell is poorly differentiated with few organelles and a significant number of free polysomes. (13,600$\times$)

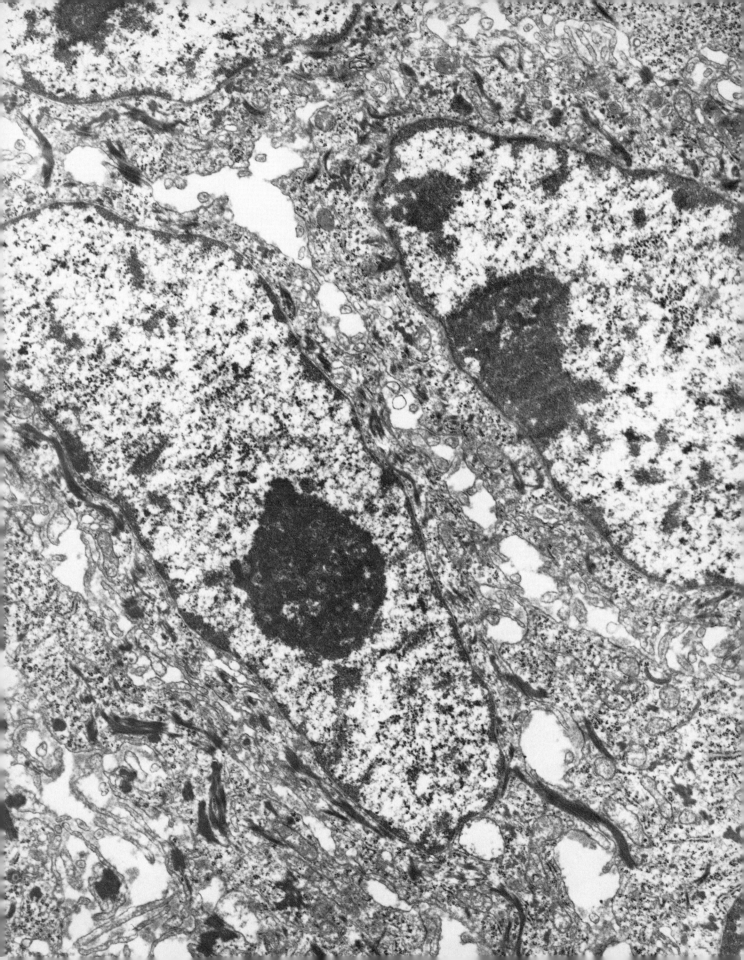

Squamous Cell Carcinoma of the Lung

FIGURE 20. There are markedly interdigitated cells with many prominent desmosomes. Tonofilaments are haphazardly arranged within the cytoplasm and are frequently attached to the desmosomes. These tonofilaments and desmosomes are characteristic of squamous epithelial cells. (22,120×)

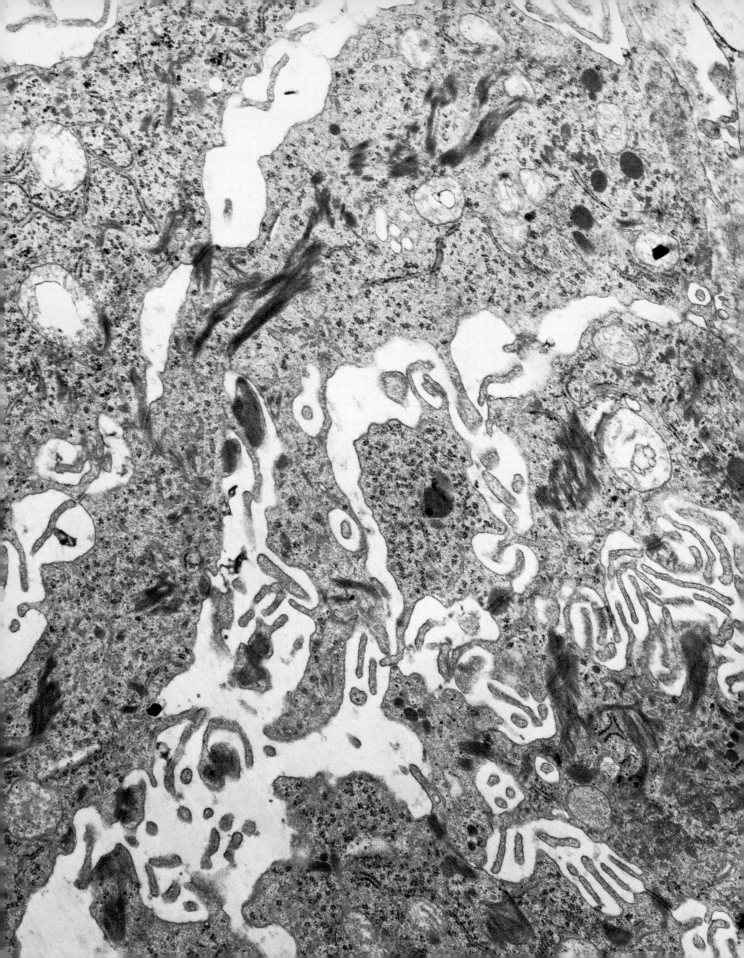

Clara Cell Carcinoma of the Lung

FIGURE 21.   This electron micrograph shows an irregular scalloped lumen of an acinus from a carcinoma of the lung. In general the neoplastic cells delineating the gland lumen are pleomorphic, contain various types of fairly well-developed organelles such as rough endoplasmic reticulum, mitochondria, Golgi complex, and glycogen. The apical portion of most of the cells forms dome or tongue-like processes with constricted zones, probably indicating an early phase of decapitation. Apocrine secretion was observed by light microscopy. This feature is characteristic of Clara cells. Clara cells are present normally in bronchial, bronchiolar epithelium, but are more abundant in the terminal bronchioles. A portion of cytoplasm is lying free in the lumen and may either represent a product of decapitation or oblique sectioning. (8600×)

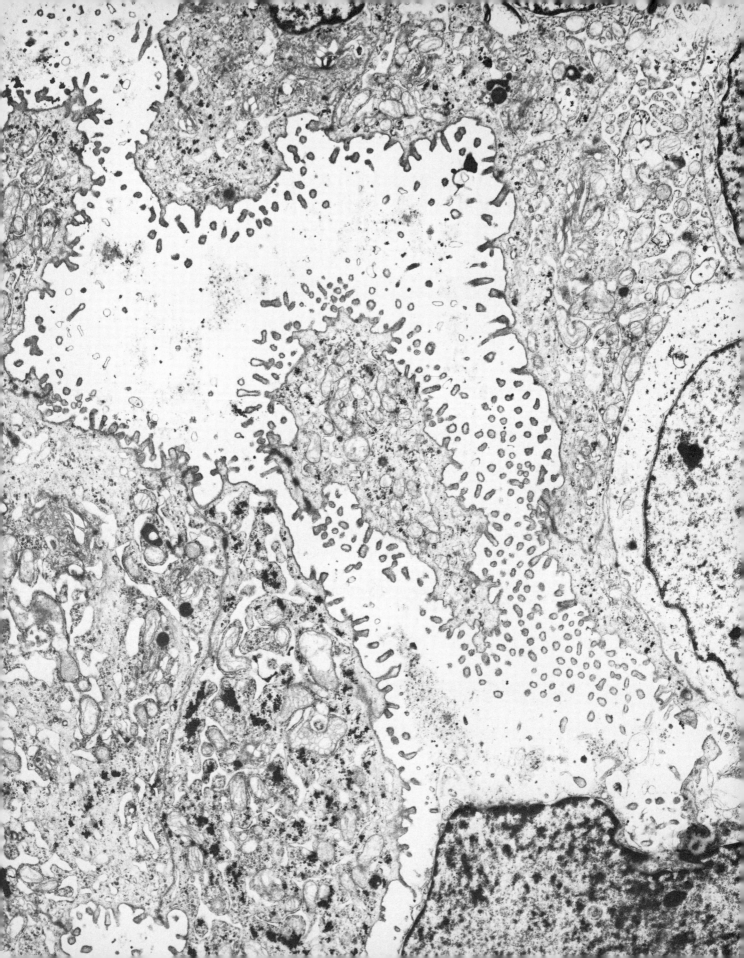

3 Parotid Gland

Adenolymphoma of the Parotid Gland

FIGURE 22. In the upper portion of the electron micrograph, the basal part of the epithelial cells are resting on a well-formed basal lamina. The stroma (lower portion of the micrograph) contains lymphocytes, a fibroblast and a mast cell. Mature collagen bundles are present between the stromal cells. The luminal portion of the epithelial cell is presented in the next figure. (10,000×)

44

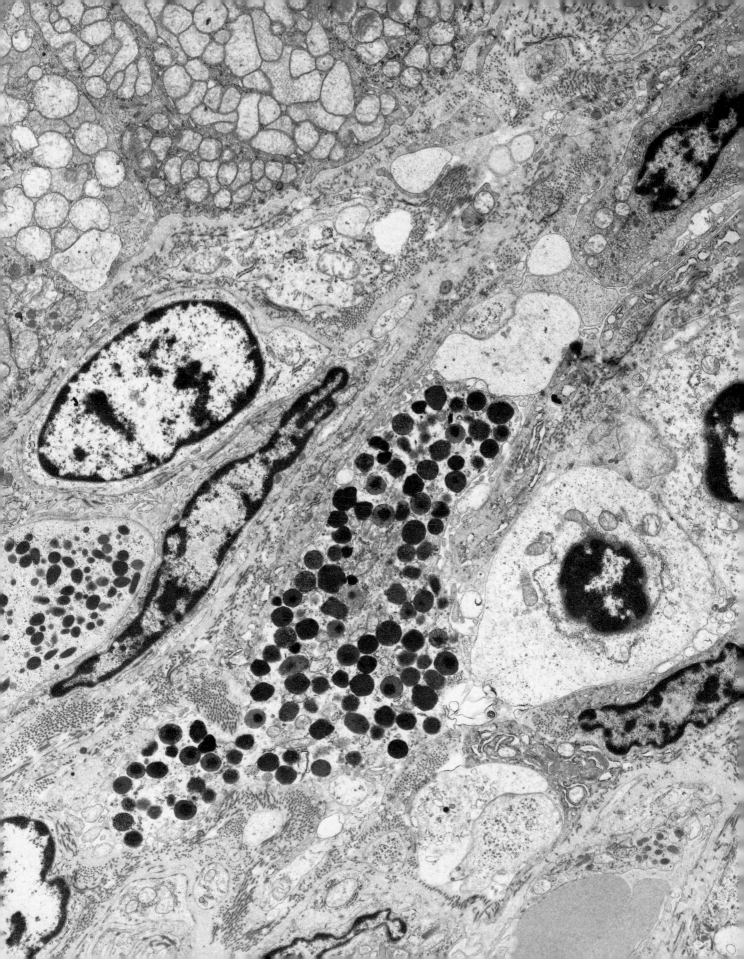

Epithelial Cells of the Adenolymphoma
Of the Parotid Gland

FIGURE 23. The apical part of the epithelial cells is in contact with a lumen. Short microvilli and a terminal web are prominent. The cytoplasm is characteristically filled with closely packed mitochondria. The cristae vary from short to elongated and concentrically arranged. The high magnification of the mitochondria is presented in the next figure. (10,000×)

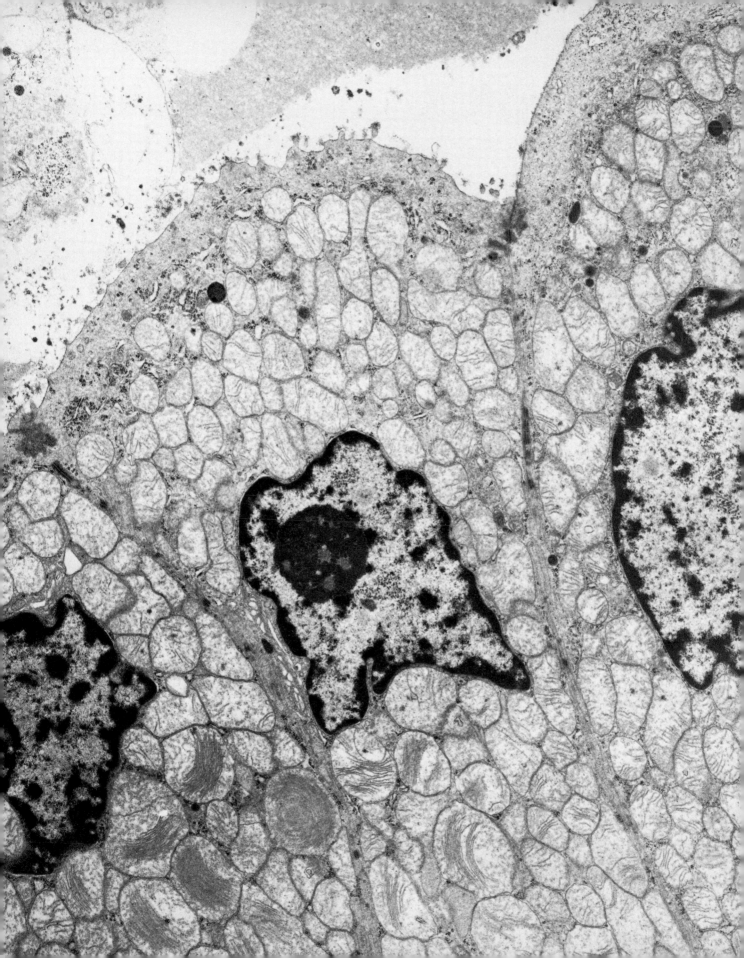

High Magnification of the Adenolymphoma
Mitochondria of the Parotid Gland

FIGURE 24.   The mitochondria show a circular or semicircular arrangement of their cristae giving a fingerprint-like appearance. The mitochondria are closely packed but occasional free ribosomes are seen between mitochondria. (53,000×)

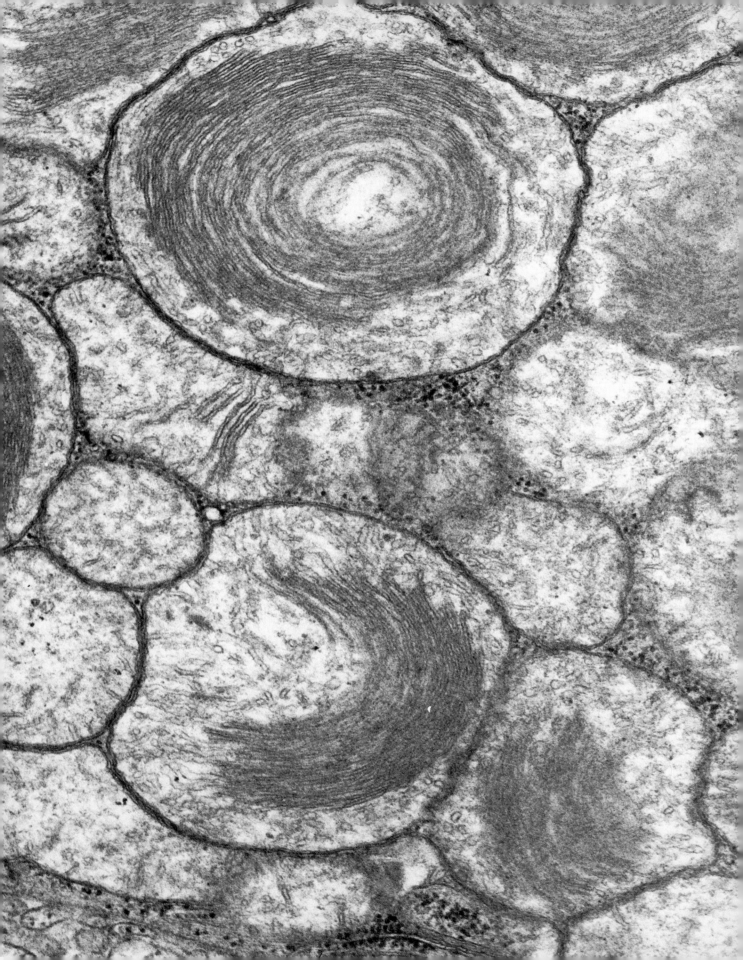

4 Intestine

Blebbing of the Plasma Membrane

FIGURE 25. This electron micrograph exemplifies a common reaction of many cells to acute cell injury, namely blebbing of the plasma membrane. The mechanism of this focal defect is unknown, but it probably is related to a loss of water and electrolyte regulation. The cells of this micrograph are small intestinal mucosal epithelial cells which normally are characterized by their striking brush border of microvilli. In this case though, there is loss and distortion of these microvilli. (34,000×)

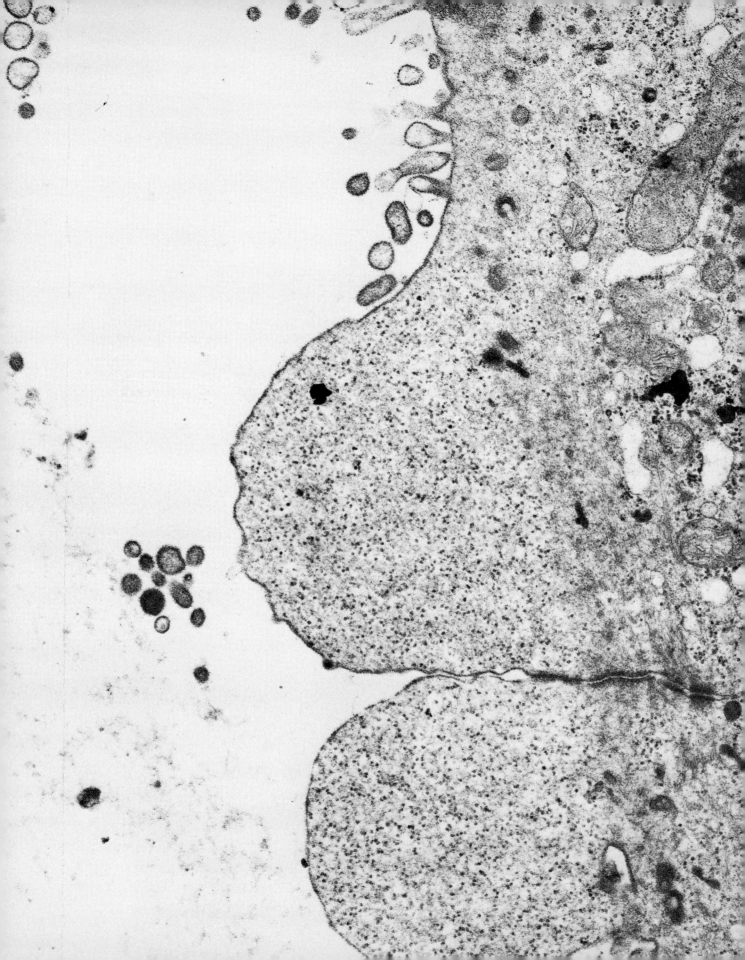

Giardia Lamblia Overlying Intestinal Epithelium

FIGURE 26. The parasite is seen in sagittal section and overlying the brush border of small intestinal epithelial cells. The parasite contains a nucleus, a prominent suction disc on its ventral surface, many flagella and endoplasmic reticulum throughout the cytoplasm. The underlying entero-cytes have distorted microvilli, swollen and damaged mitochondria, and endoplasmic reticulum. ($22,600\times$)

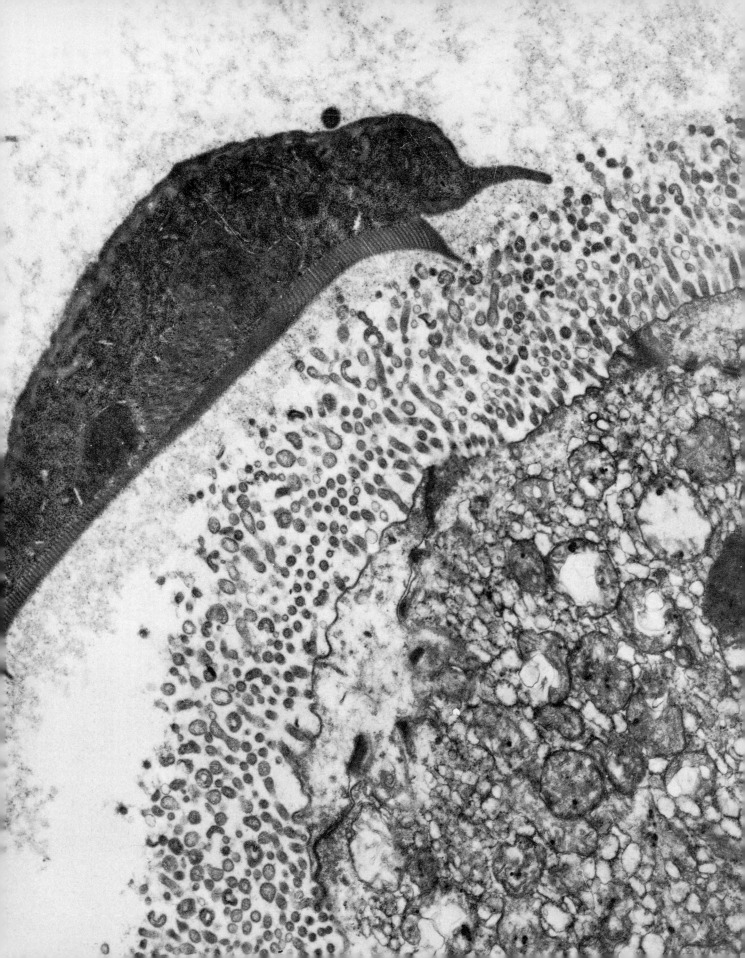

Intestinal Epithelium in Celiac Disease

FIGURE 27.   The structural changes seen in enterocytes are short, stubby, and often branching microvilli, an indistinct terminal web, an absent glycocalyx (fuzzy coat), and the presence of lysosomes in the cytoplasm. (11,600×)

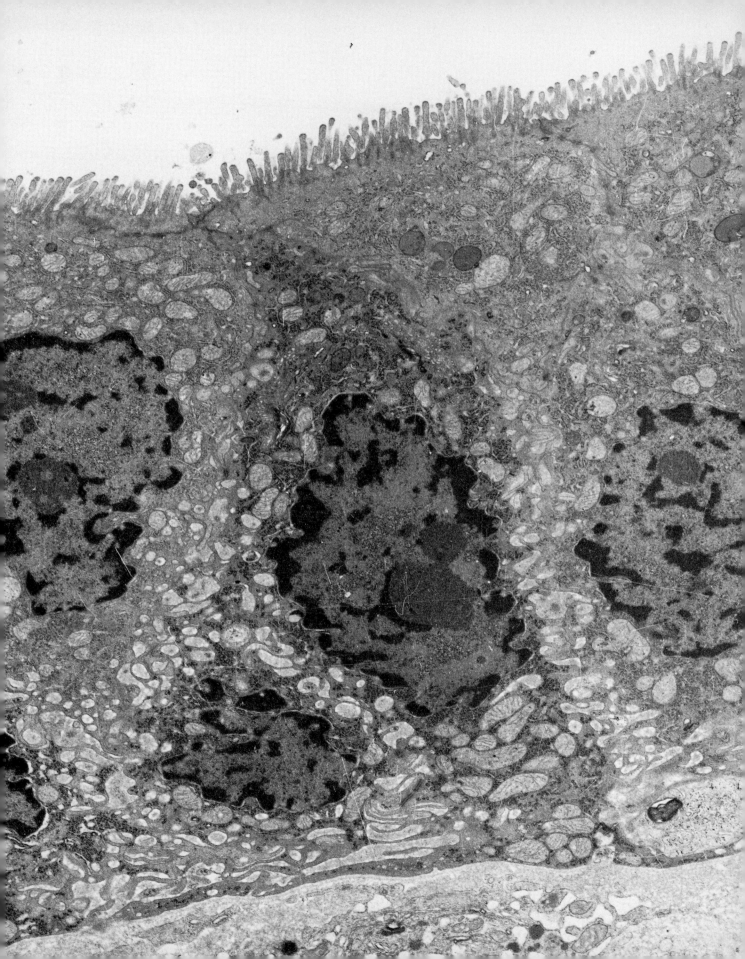

Leiomyosarcoma of the Jejunum

FIGURE 28. The tumor cells are elongated and have elongated nuclei with well-defined nuclear envelopes delineating a tortuous contour. The nuclear chromatin is granular and is condensed along the nuclear border. Intracytoplasmic organelles are abundant, and mitochondria appear to be located at the nuclear poles. The most striking feature is the presence of myofilaments organized in globular clumps. (18,700×)

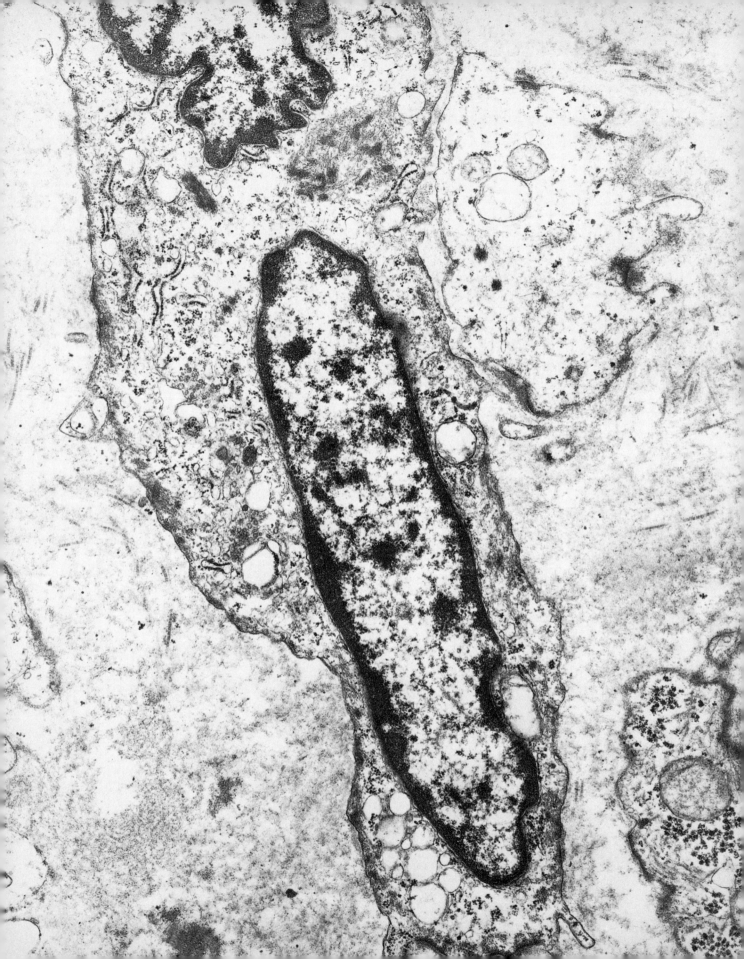

Leiomyosarcoma of the Jejunum

FIGURE 29.   The tumor cells have a distinct basal lamina and irregular densities on the inner portion of the cell membrane. The periphery of the cell cytoplasm is richly endowed with well-defined pinocytotic vesicles. Abundant myofilaments are seen within the cytoplasm. ($20,900\times$)

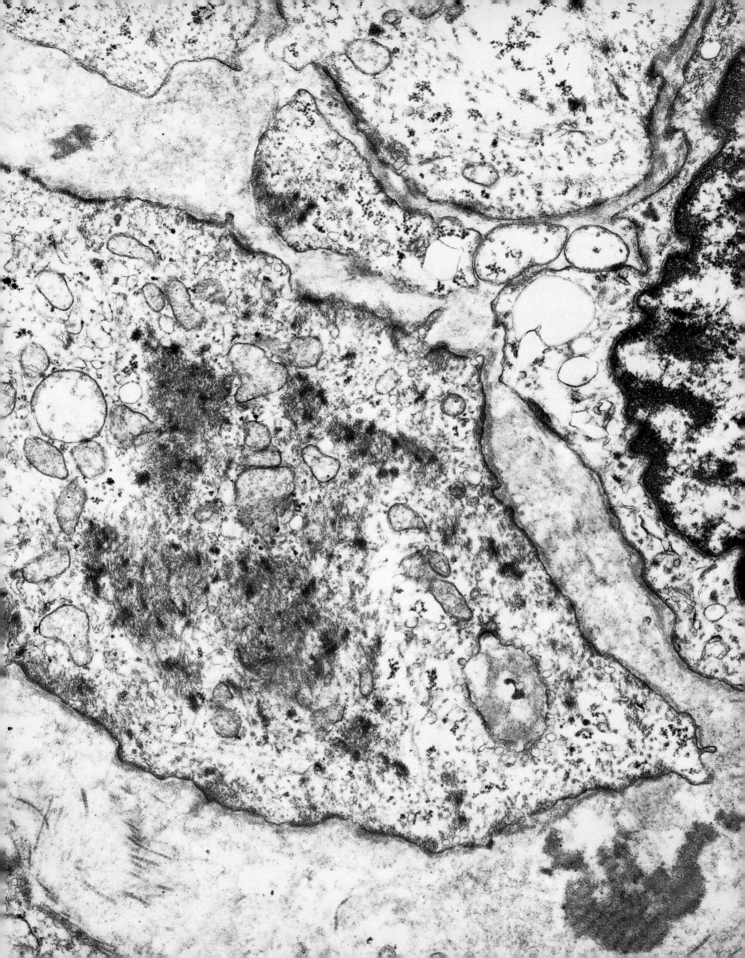

Myofilaments in Leiomyosarcoma

FIGURE 30. Abundant myofilaments are seen in the cytoplasm. They are organized in elongated bundles, measure 8 to 9 nm in thickness and seem to concentrate at irregular intervals forming characteristic dense bodies. (37,100×)

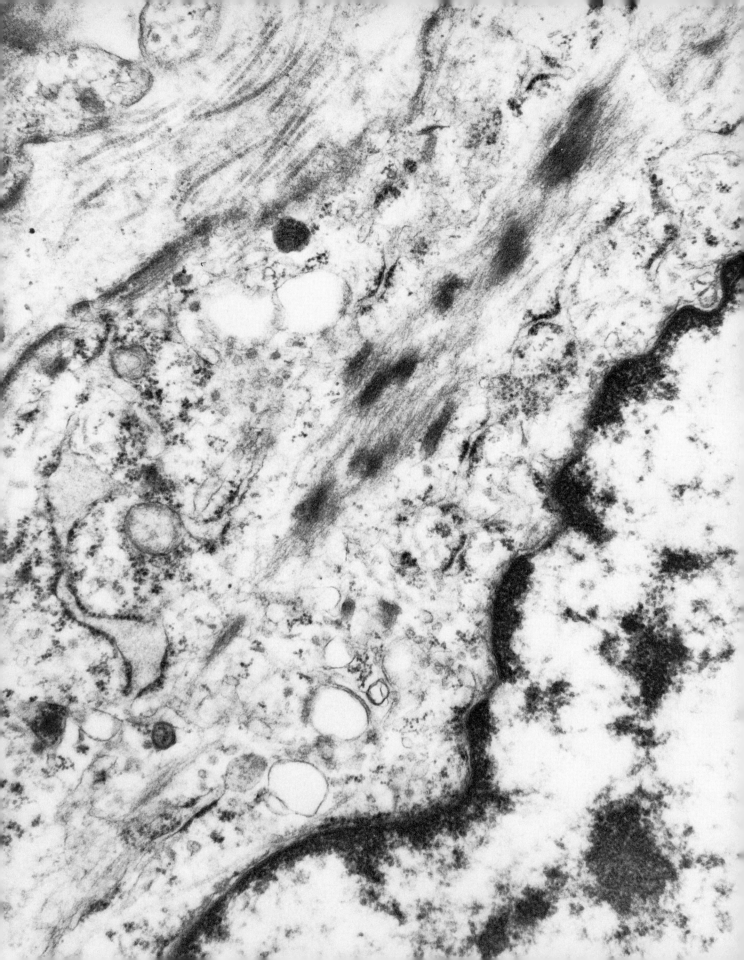

Carcinoid Tumor of the Ileum

FIGURE 31. A portion of the nucleus and adjacent cytoplasm from a carcinoid tumor cell discloses numerous pleomorphic, membrane-bound secretory granules (enterochromaffin granules) of a variable electron density with a clear halo of nearly uniform width around the granule core. The granules are mainly distributed in the basal cytoplasm. They measure an average 230 nm with a range of 100 to 360 nm. Carcinoid tumors arise from the enterochromaffin cells which are dispersed throughout various organs, predominantly in the crypts of Lieberkühn, and are found in various epithelia such as the epithelium of the gastrointestinal system, gallbladder, bile and pancreatic ducts, trachea, bronchi, uterine cervix as well as the thymus, breasts, and gonads. (45,000×)

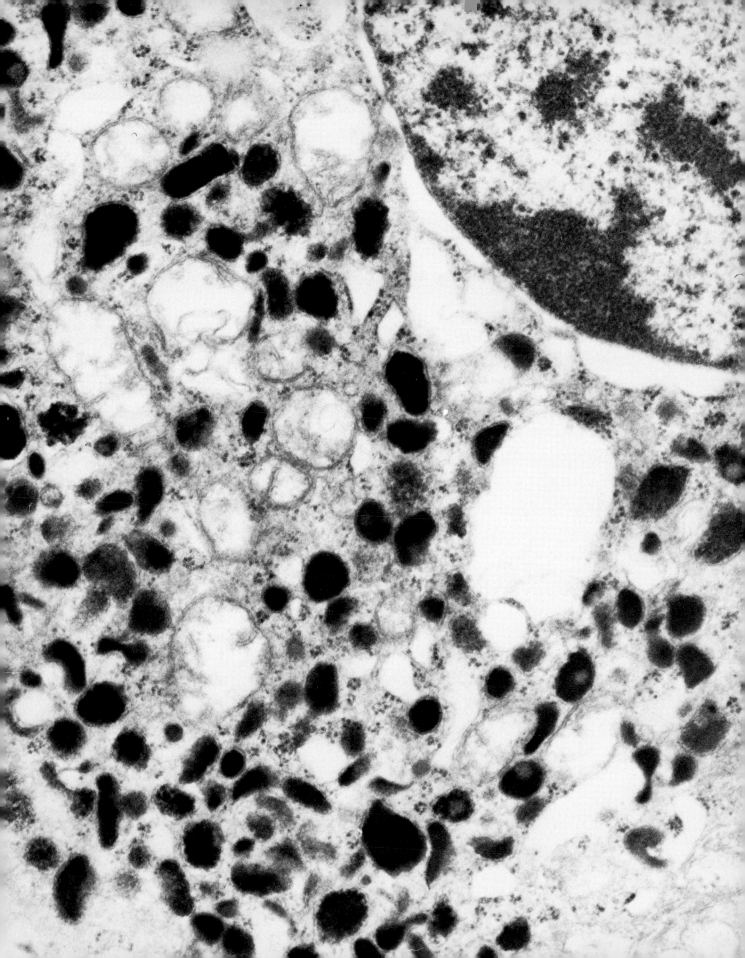

Mucin-Producing Adenocarcinoma of the Colon

FIGURE 32. Pseudostratified columnar epithelium extends from a delicate basal lamina to the gland lumen which shows a few microvilli. The outstanding feature is the presence in the apical cytoplasm of coalescent vacuoles of irregular shape and size. They appear to be associated with the rough endoplasmic reticulum. These vacuoles contain flocculent material. A similar material is observed in the rough endoplasmic reticulum, probably indicating the site of synthesis. (11,000×)

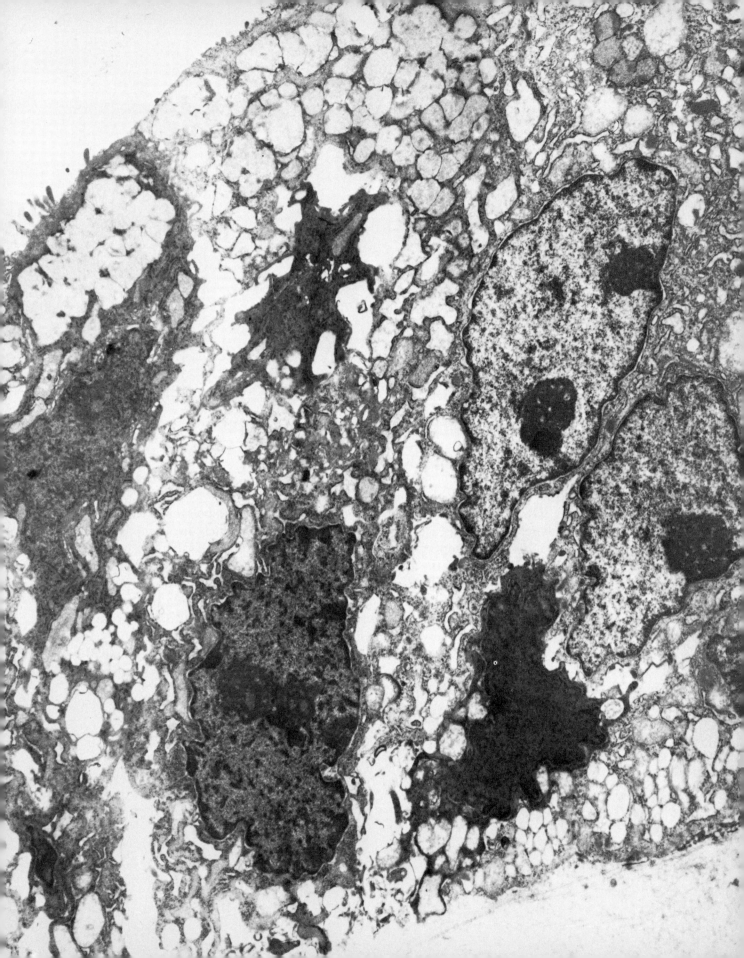

5 Liver

Hepatic Nuclear Glycogen

FIGURE 33.   The euchromatic portion of the nucleoplasm of this hepatic nucleus contains a myriad of glycogen particles which are normally found only in the cytoplasm. The nuclear glycogen is present as monoparticulate, discrete units of the beta glycogen type normally not found in the hepatocyte. This unusual type of glycogen can be contrasted to the normal cytoplasmic alpha glycogen (closed arrow) situated just outside the nucleus in this figure. Ribosomes (open arrow) being particulate are often confused with glycogen but can be easily contrasted here. Nuclear glycogen is most often reported to occur in hepatocytes from patients with diabetes mellitus although it has also been reported to be associated with many nondiabetic conditions. (31,000×)

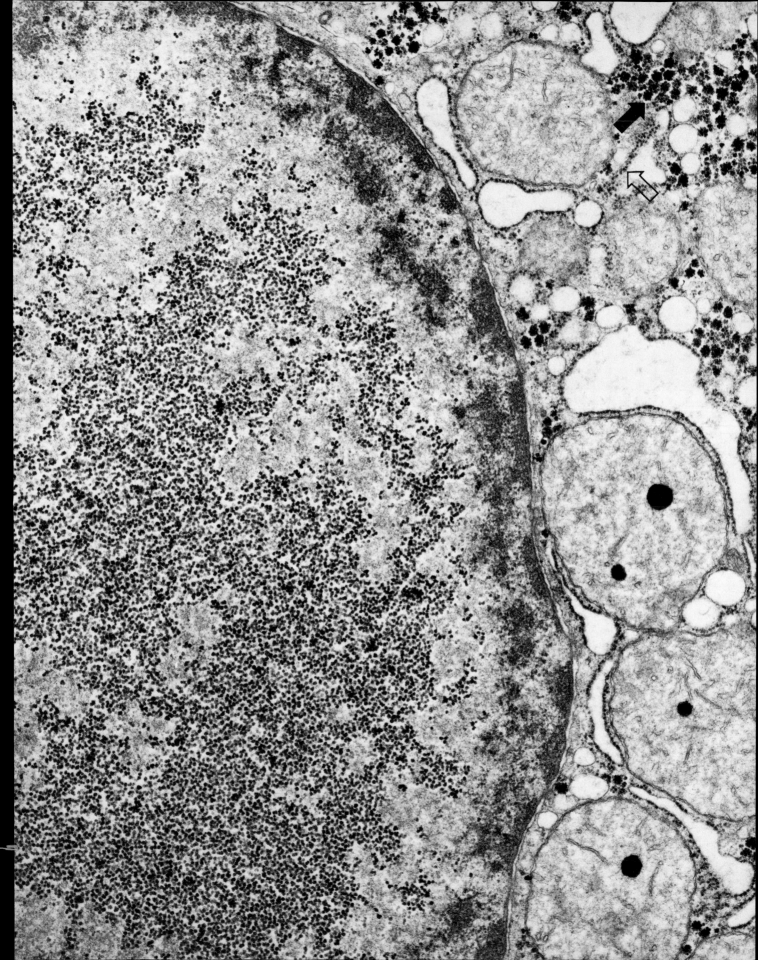

Hepatitis B Core Antigen in a Hepatocyte Nucleus

FIGURE 34.   We have encircled some of the many small particles, singly and in aggregates, that have an open circle structure in the nucleoplasm of a hepatocyte. A better appreciation of the structure of these particles can be obtained from the high magnification in the next figure. Such particles are consistently 20 to 25 nm in diameter and have been associated with the core antigen of the hepatitis B virus. These nuclear structures and cytoplasmic surface antigen structures which are shown in another figure, unite, possibly along with other yet unknown components, to form the infectious hepatitis B virus that has been isolated from serum as the Dane particle. (32,000×)

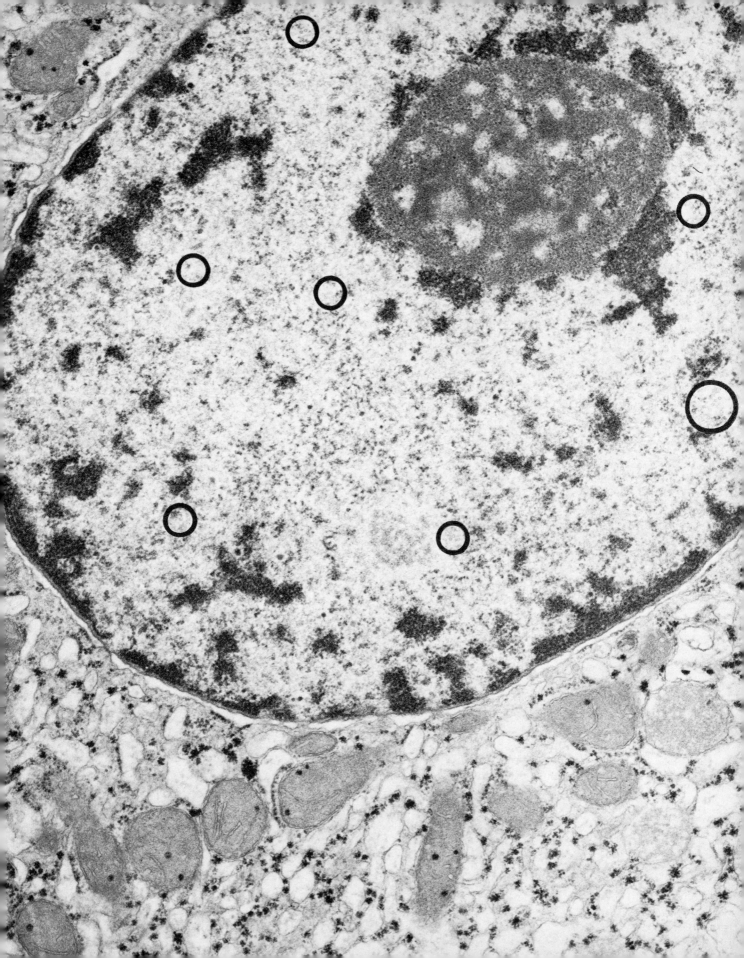

Higher Magnification of Hepatitis B
Core Particles in a Hepatocyte Nucleus

FIGURE 35.   In this micrograph of hepatocyte nucleoplasm, two hepatitis B core particles are encircled. Many other particles are present, some singly, others in groups or aggregates. (82,000×)

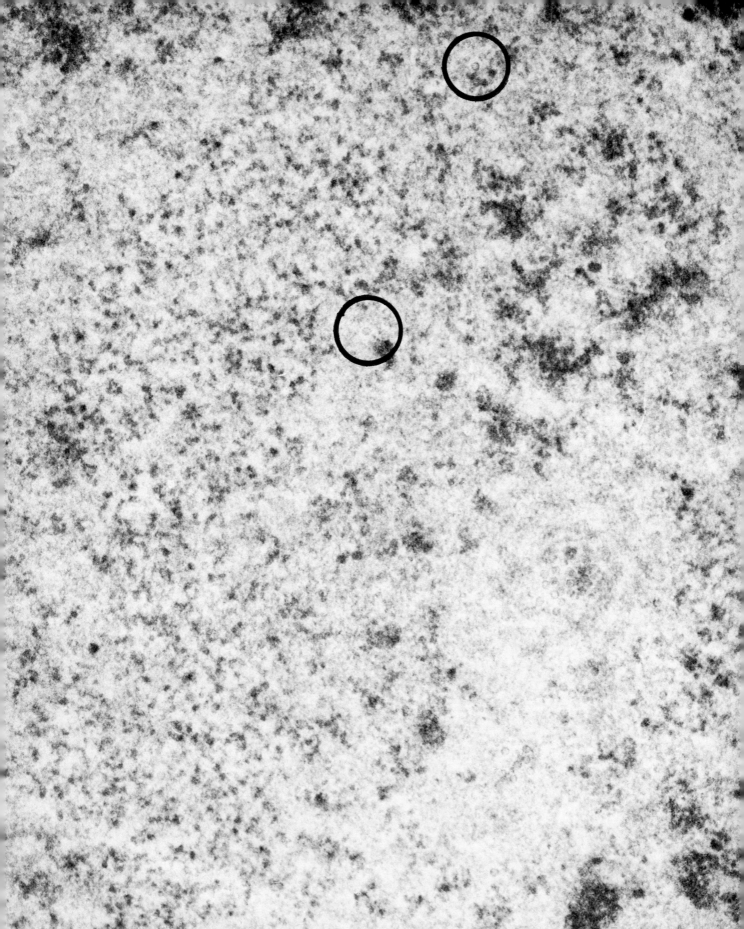

Hepatitis B Surface Antigen in Hepatocyte Cytoplasm

FIGURE 36. The smooth endoplasmic reticulum (SER) in the hepatocytes in this electron micrograph is markedly hyperplastic. More exceptional is the occurrence of filamentous and particulate material in the usually clear luminal space. The material while not distinct at the magnification of the figure does give a texture characteristic of hepatitis B surface antigen, especially in the hepatocyte on the left although some is present in the hepatocyte on the right. A more distinct view is furnished in the higher magnification of the next micrograph. The surface antigen, formerly known as Australian antigen, is considered a component on the surface of the hepatitis B virus. (17,000×)

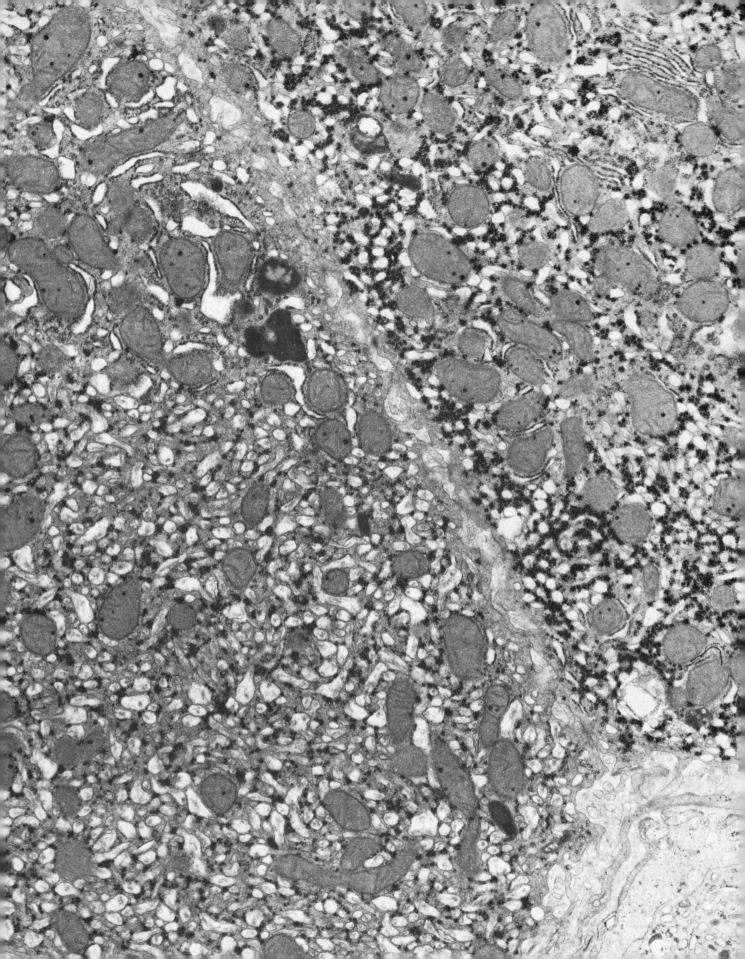

Higher Magnification of Hepatitis B Surface Antigen

FIGURE 37. Both the filamentous, longitudinal aspect (long arrow) and the circular, cross-sectional aspect (short arrow) of the hepatitis B surface antigen particles can be appreciated in this high resolution electron micrograph. These particles within the lumina of the smooth endoplasmic reticula (SER) give the characteristic texture noted in the previous lower magnification electron micrograph. (54,000×)

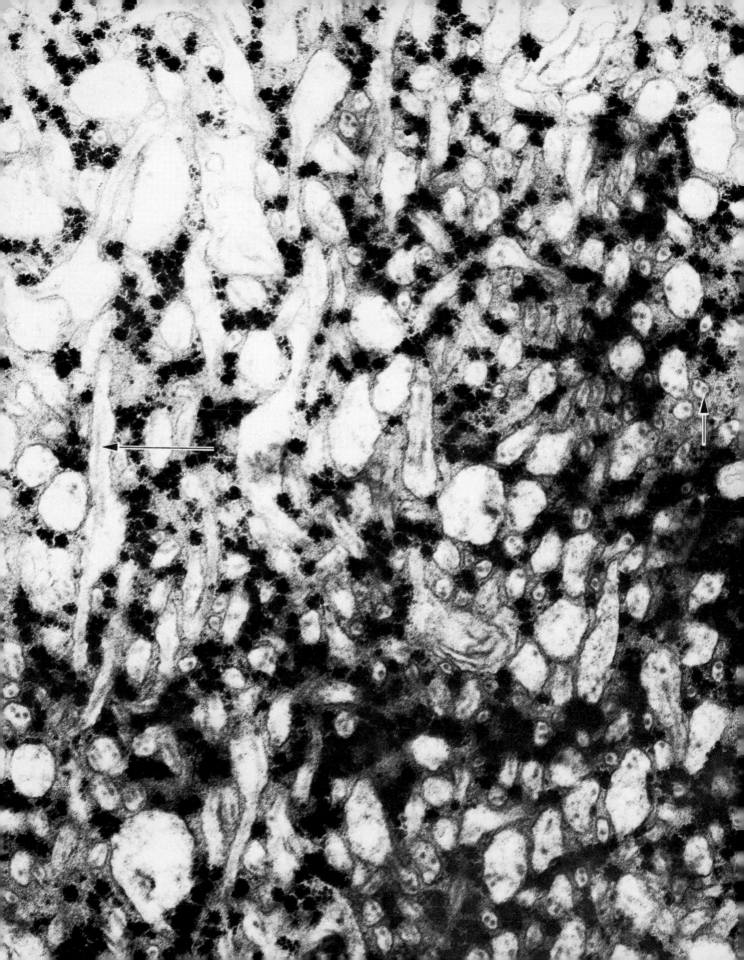

Acidophilic Body of the Liver

FIGURE 38.   The so-called acidophilic body, distinguished by its intense eosinophilia under light microscopy, appears as a shrunken, necrotic hepatocyte under electron microscopy. The nucleus, when present, is pyknotic, the mitochondria are swollen, and the cytoplasmic matrix is exceptionally dense. An easy comparison is available with the normal hepatocyte at the bottom of the figure. Phagocytosis by the sinusoidal Kupffer cell occurs and shows well in the next figure. Acidophilic bodies are characteristically found in viral and drug hepatitides although not exclusively. (12,000×)

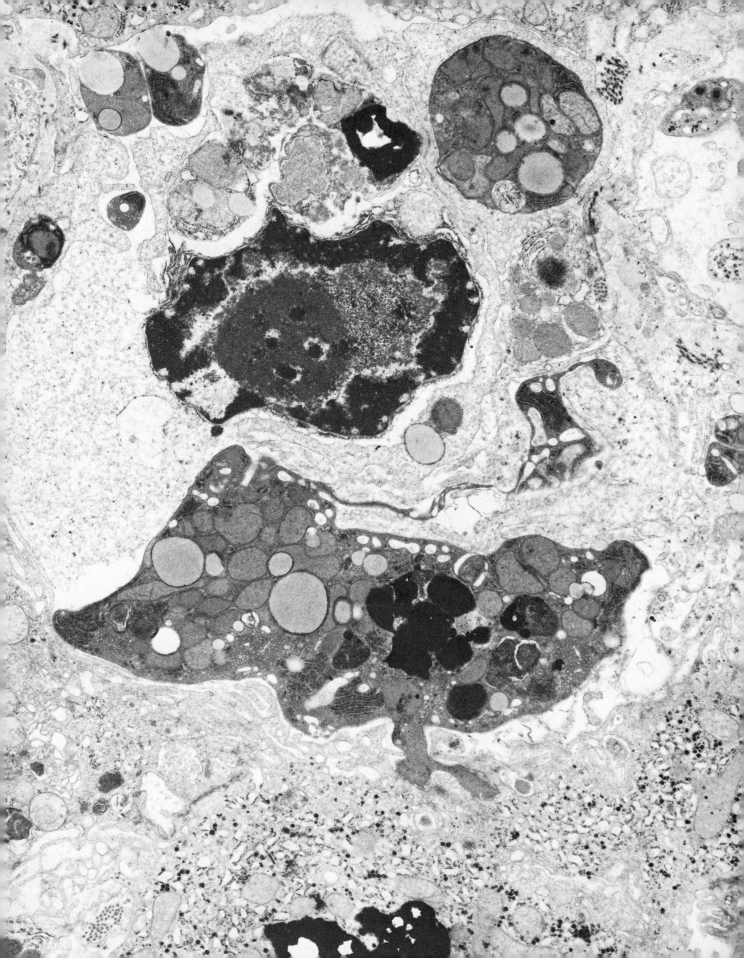

Acidophilic Body Phagocytosed by a Kupffer Cell

FIGURE 39.   The sinusoidal Kupffer cell contains a round necrotic hepatocyte with a pyknotic nucleus and very dense cytoplasm. The dense cytoplasm causes the intense eosinophilia or acidophilia that characterizes these bodies by light microscopy. The Kupffer cell will digest the phagocytosed cell until only an electron dense residual body (arrow) remains. The acidophilic body is first apparent at the site of a hepatocyte but subsequently is expelled to the sinusoid (see the preceding figure) and finally phagocytosed as shown here. (19,000×)

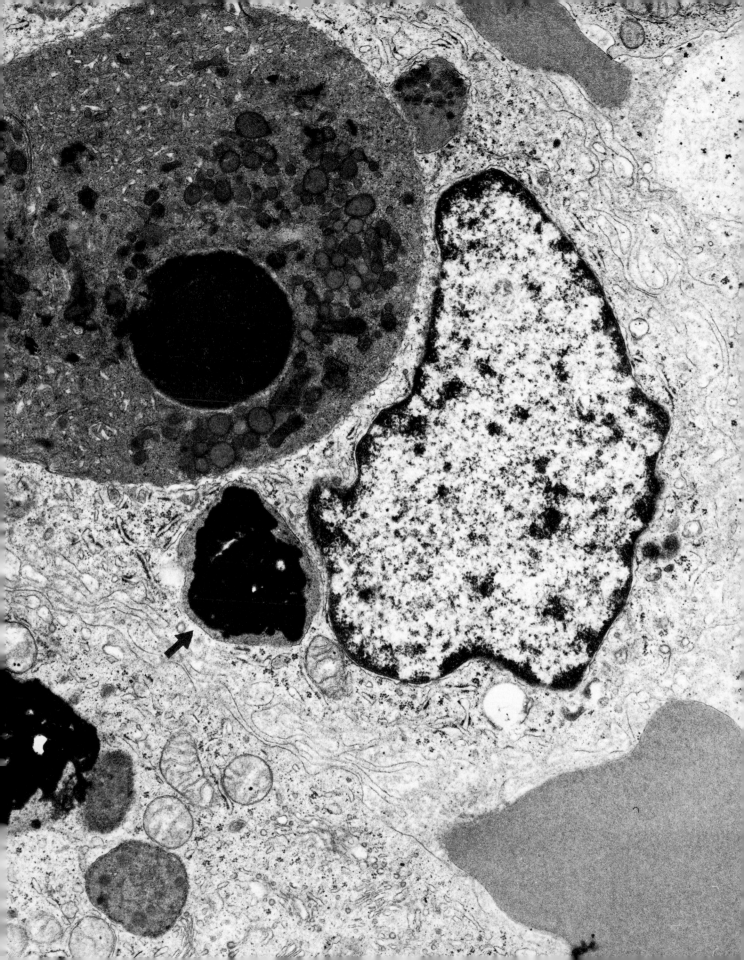

Acute Cellular Injury

FIGURE 40. This abnormal hepatocyte exemplifies some of the ultra-structural changes associated with acute cellular injury. The mitochondria and endoplasmic reticulum are swollen. The light mitochondrial matrices and endoplasmic reticulum lumina suggest that the swelling is due to water. The angular shape of the cell and the dense nucleoplasm are also signs of injury. (15,000×)

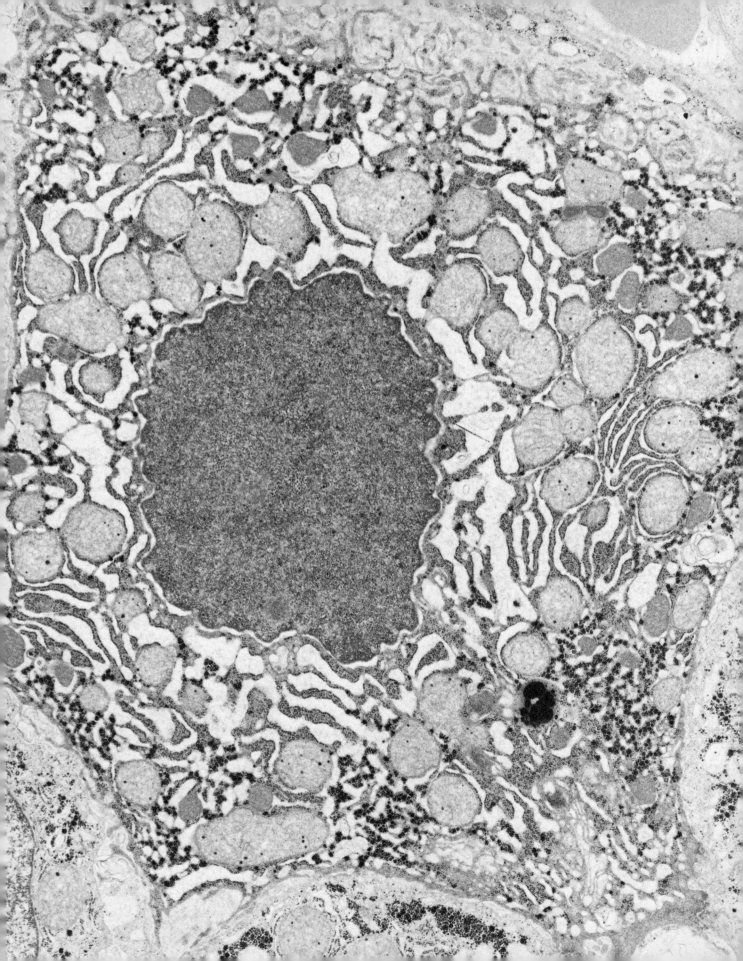

Necrotic, Binucleated Hepatocytes

FIGURE 41. The nucleoplasm of the two nuclei in this one hepatocyte is rarefied from karyolysis which is a classical sign of necrosis. In the cytoplasm the mitochondria show signs of irreversible cell injury that leads to necrosis. The mitochondria are swollen and their matrix granules are diffusely enlarged. (20,000×)

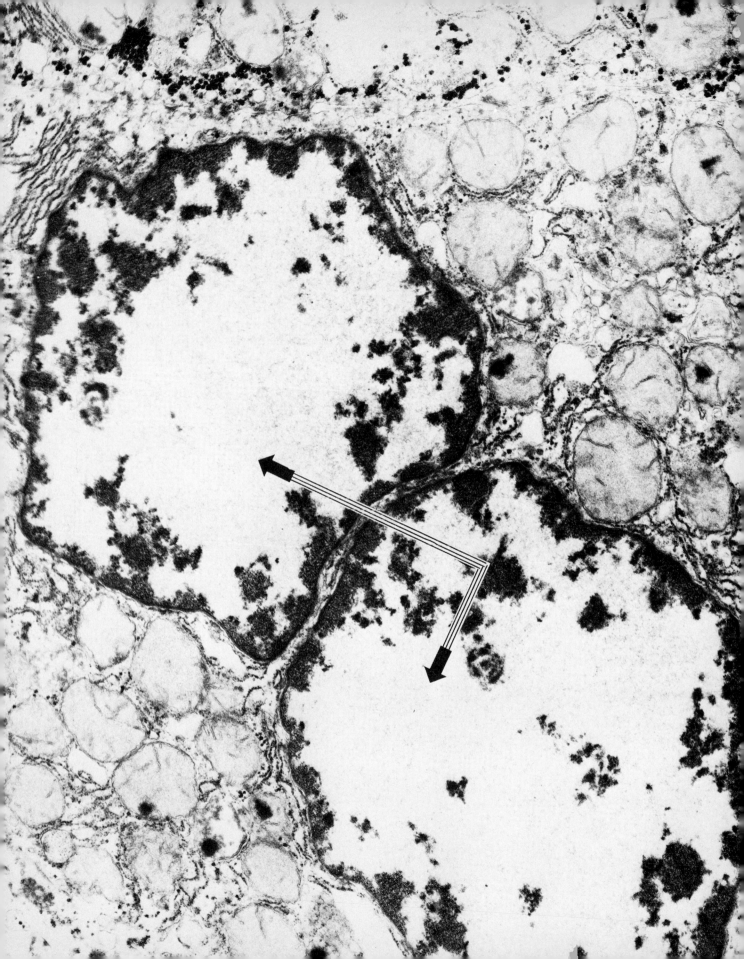

Extracellular Cell Fragments

FIGURE 42. An artifact often encountered in electron microscopy of human biopsy specimens like liver is the occurrence of small cell fragments displaced from the other usual cellular components. These fragments are usually delimited by a plasma membrane as in this electron micrograph from a liver biopsy specimen. Other times the cytoplasmic organelles are free in the extracellular space. (23,000×)

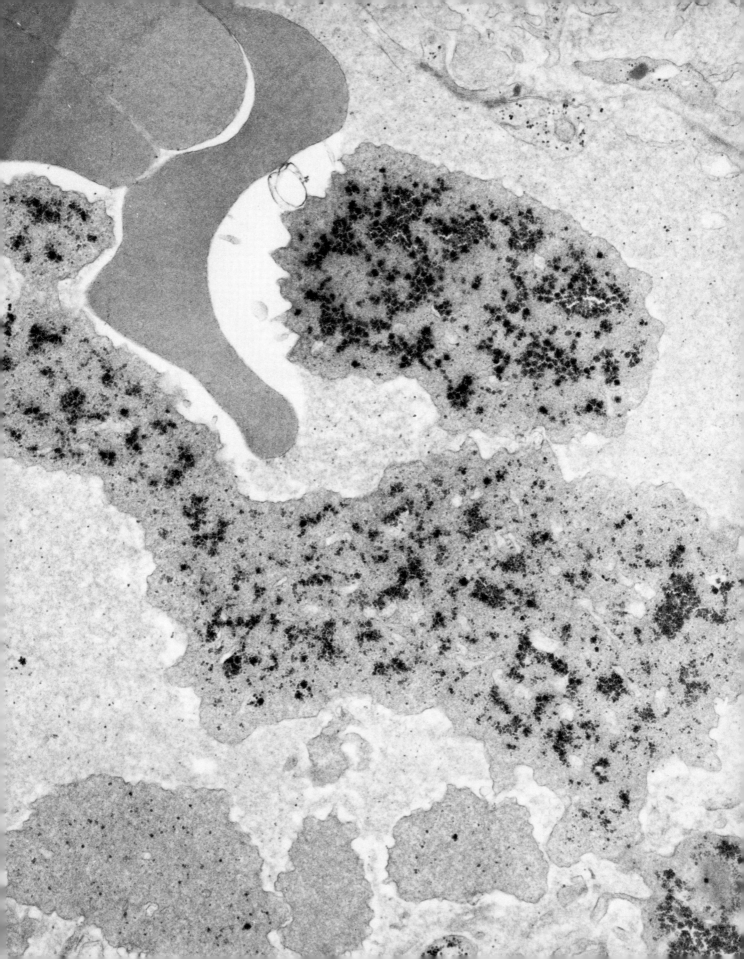

Giant Mitochondrion with Filaments in a Hepatocyte

FIGURE 43.   The abnormally large structure to the right of the hepatocyte nucleus is identified as a mitochondrion by these normal features: (1) its double membrane; (2) cristae; and (3) matrix granules. The numerous patches of filaments are abnormal but found often in hepatocytes in liver disease as well as in many other cell types. Giant mitochondria or mitochondrial filaments are usually, but not necessarily, associated with alcoholic liver disease, and as demonstrated in the next two figures are not always found together. (35,000×)

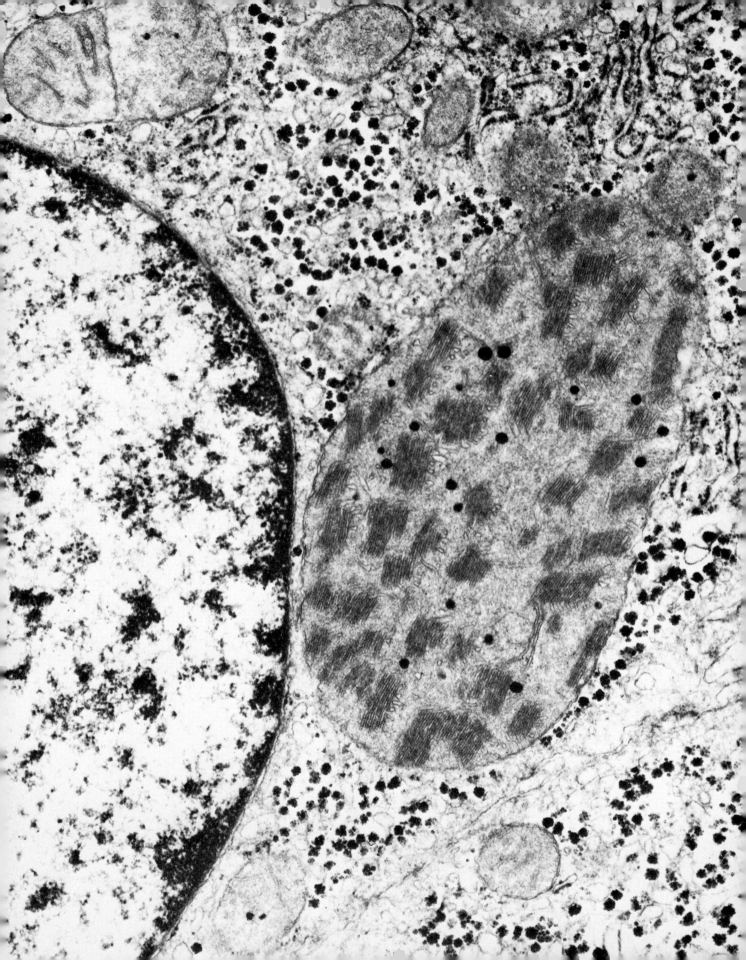

Giant Mitochondrion (Hepatocyte)

FIGURE 44.   The abnormally large size of the one mitochondrion is easily appreciated in comparison to the more normal, nearby mitochondria. The giant mitochondrion lacks all but one very long crista that is aligned parallel to and immediately under the double membrane of the mitochondrion. Giant mitochondria are easily resolved by light microscopy although unequivocal identification as mitochondria must be made by electron microscopy. As demonstrated in the previous figure, giant mitochondria at times abnormally contain filaments. (50,000×)

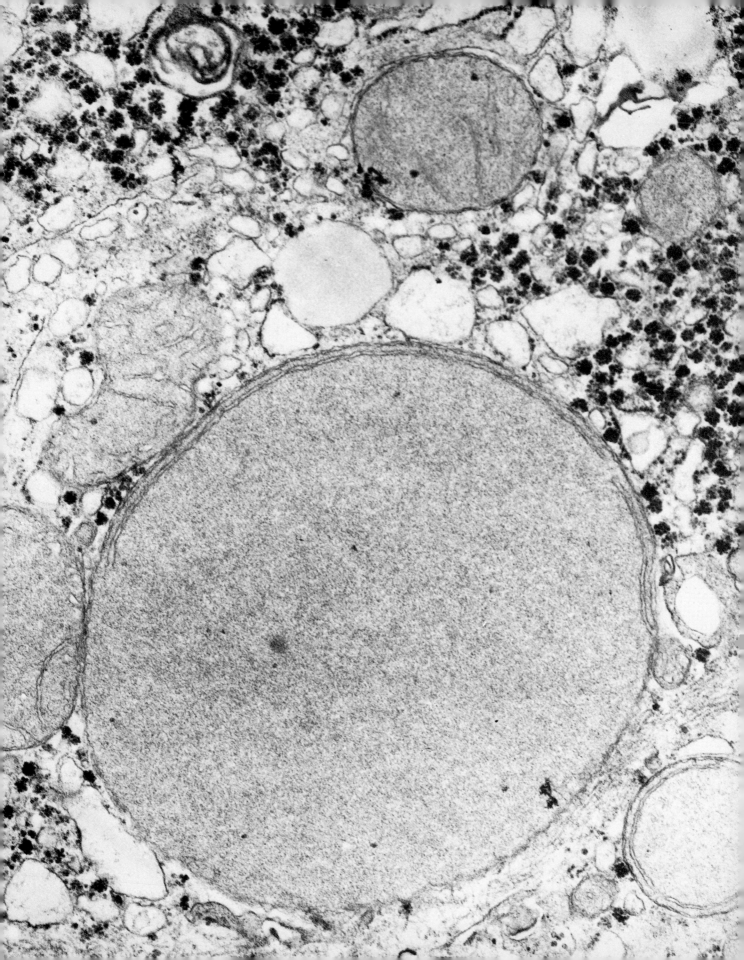

Mitochondrial Filaments

FIGURE 45.   Most of the hepatic mitochondria in this electron micrograph show patches of very straight, long, parallel filaments. Although filaments are not considered a normal mitochondrial feature, their occurrence has been reported in various animals, tissues, and diseases. Mitochondrial filaments are also found in giant mitochondria as demonstrated in a preceding figure. The nature of the filament is essentially unknown. (43,000$\times$)

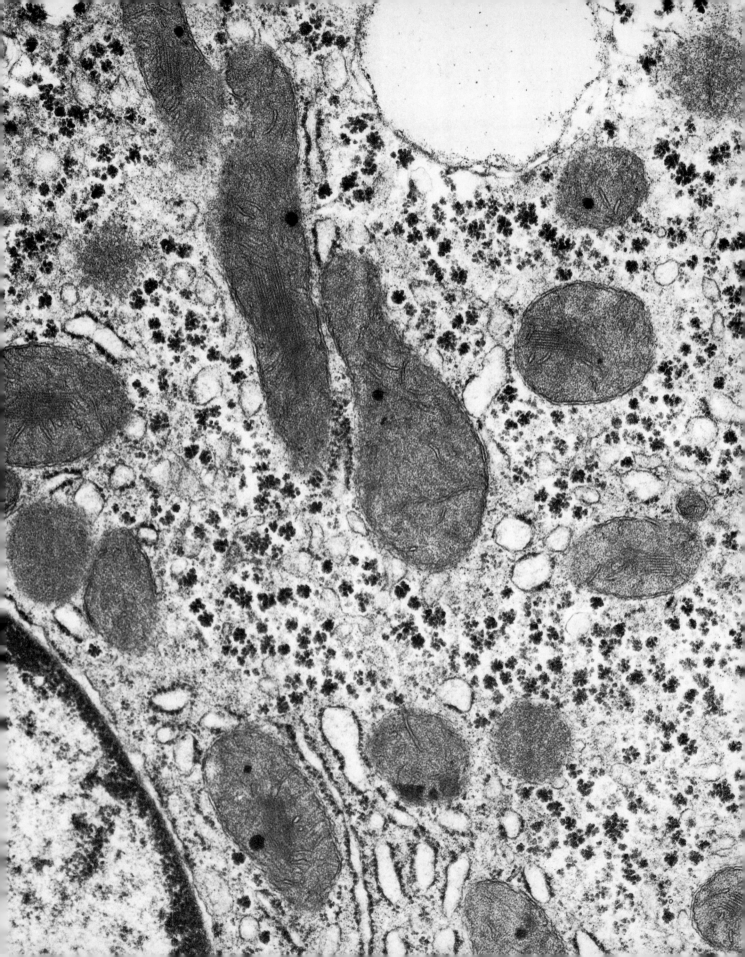

Cholestatic Features of Hepatocytes

FIGURE 46. Three features found in hepatocytes during cholestasis are presented in this micrograph.

1. The bile canaliculus (arrow) is marked by distortion and loss of normal microvilli.
2. The pericanalicular ectoplasm that is organelle-free is widened.
3. There is pericanalicular, electron-dense, myelin-like material probably representing intracellular bile or focal cytoplasmic degradation effected by intracellular bile.

Other features like curled cristae and cholesterol crystals are presented in the next figure. (21,000×)

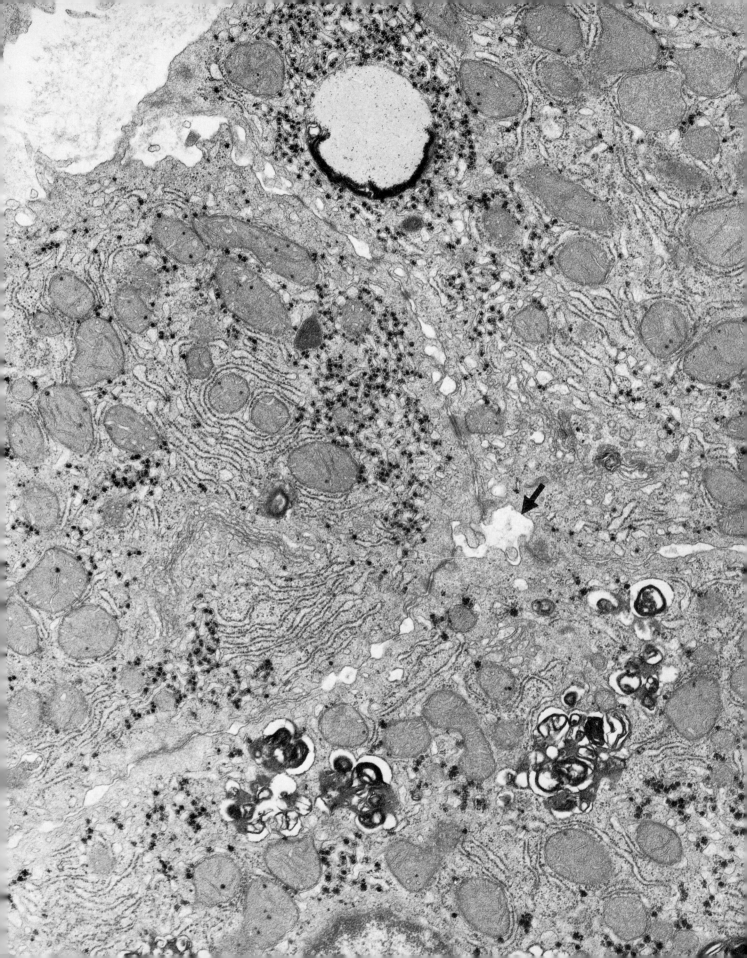

Liver

Curled Cristae of Hepatic Mitochondria
And a Lysosomal Cholesterol Crystal

FIGURE 47. The rectangular, electron transparent area in the micrograph represents a cholesterol crystal extracted during tissue processing. The crystal is in a single membrane-bound dense body presumably a secondary lysosome or phagosome. Most of the other less dense bodies are double membrane-bound mitochondria with abnormally "curled" cristae instead of the normal relatively straight cristae.

The cholesterol crystals and curled mitochondrial cristae are some of the ultrastructural features in cholestatic liver disease. Other features are presented in the preceding figure. (39,000×)

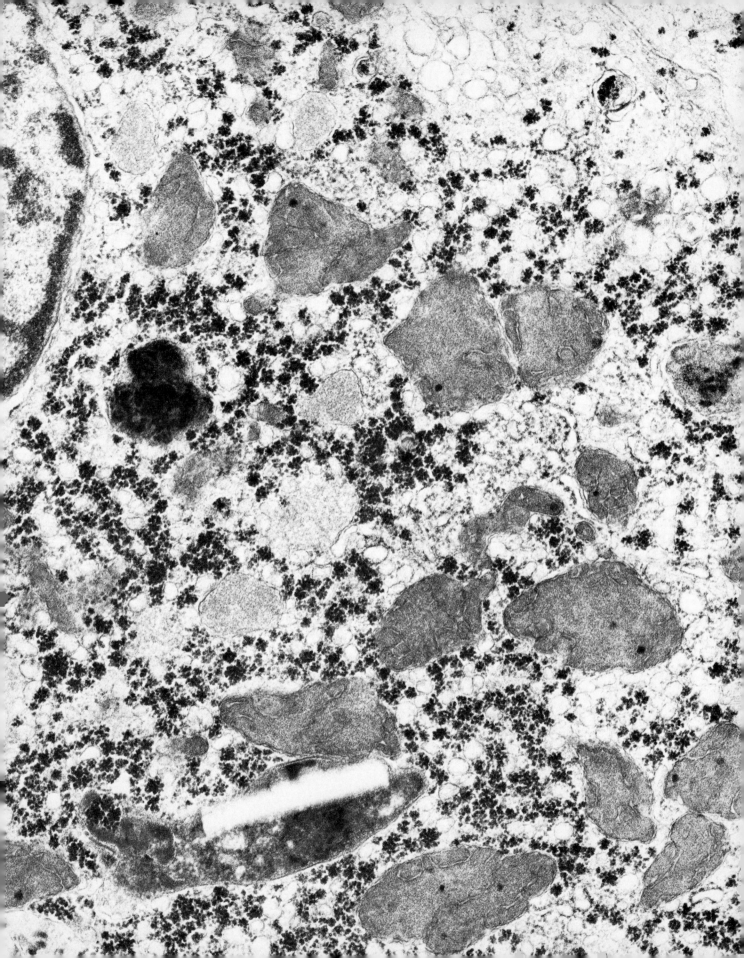

Cholesterol Crystals and Abnormally Large Mitochondrial Matrix Granules in a Hepatocyte

FIGURE 48.   The electron transparent, rectangular areas to the left side of the micrograph are thought to represent the crystals of cholesterol that have been extracted during the processing of the tissue for electron microscopy. Cholesterol crystals are often associated with high serum cholesterol levels. Also present in the micrograph are abnormally large, very electron-dense granules in the matrix of many mitochondria. These features are from a hepatocyte of a patient with alcoholic liver disease, but the features, although abnormal, are not diagnostic. ($22,600\times$)

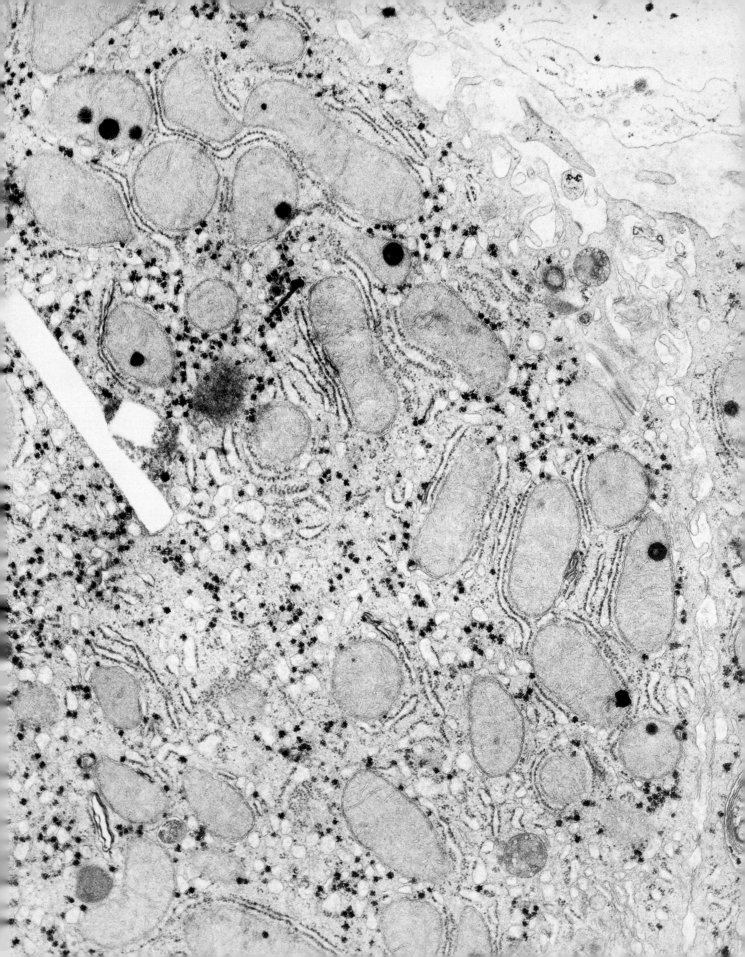

Mallory Body (Parallel, Whorled Fibrils)

FIGURE 49. The fibrillar material in this electron micrograph is the ultra-structural correlate of the alcoholic hyalin or Mallory body recognized by light microscopy in hepatocytes typically from chronic alcoholics but also from a few other liver conditions. The material is characterized by fibrils that are on the average 14 to 15 nm thick and that are not bounded by any membrane. The fibrils are found arranged in two different ways: parallel, whorled array as in this figure, or randomly interwoven array as in the next figure.

The electron translucent, globular bodies are fat droplets. A nucleus is situated in the left part of the figure. (40,000×)

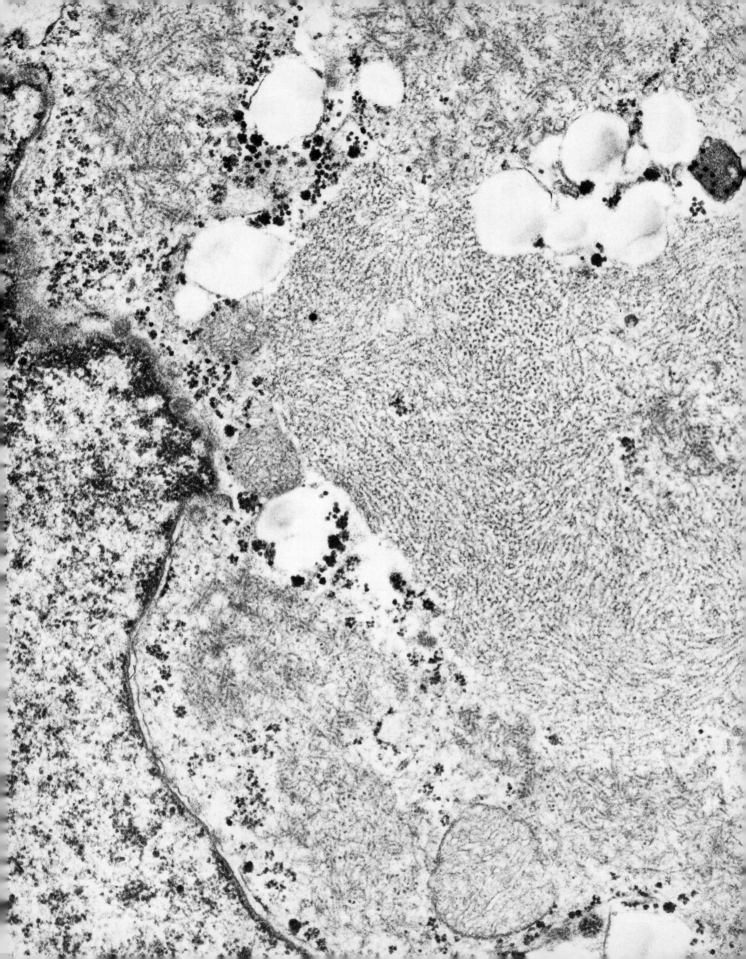

Mallory Body (Random Fibrils)

FIGURE 50. The fibrils in this electron micrograph from a hepatocyte are nonmembrane bound and randomly arranged in a felt-like interwoven array. The material is the ultrastructural correlate of one type of Mallory body. The other ultrastructural type is demonstrated in the preceding figure. (40,000×)

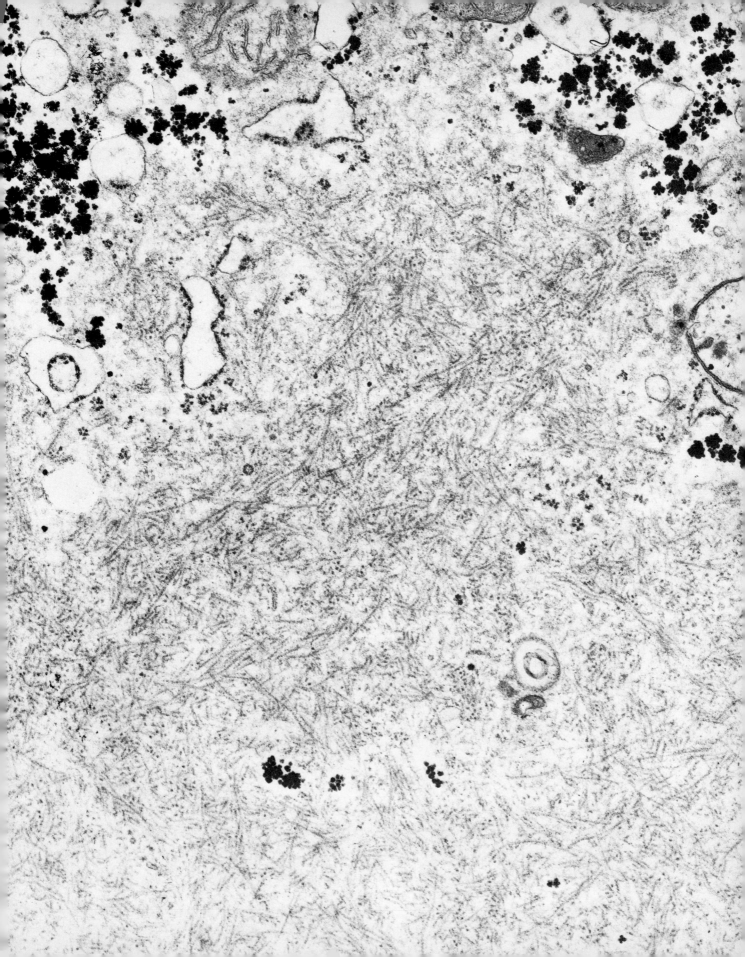

Liver

Fatty Liver

FIGURE 51. Fat accumulation is a common ultrastructural effect of numerous agents that injure the liver. A common example of such an agent is alcohol. The fat accumulates as variously sized, round drops. The electron density of the fat depends on the processing procedure and stains used and possibly the kind of fat (see Fat Storing Cell, Figure 71). One of numerous mechanisms proposed for the fat accumulation is that of an agent-induced defect in the protein synthetic part of lipoprotein synthesis. The lipid is not utilized, and therefore accumulates. (11,700$\times$)

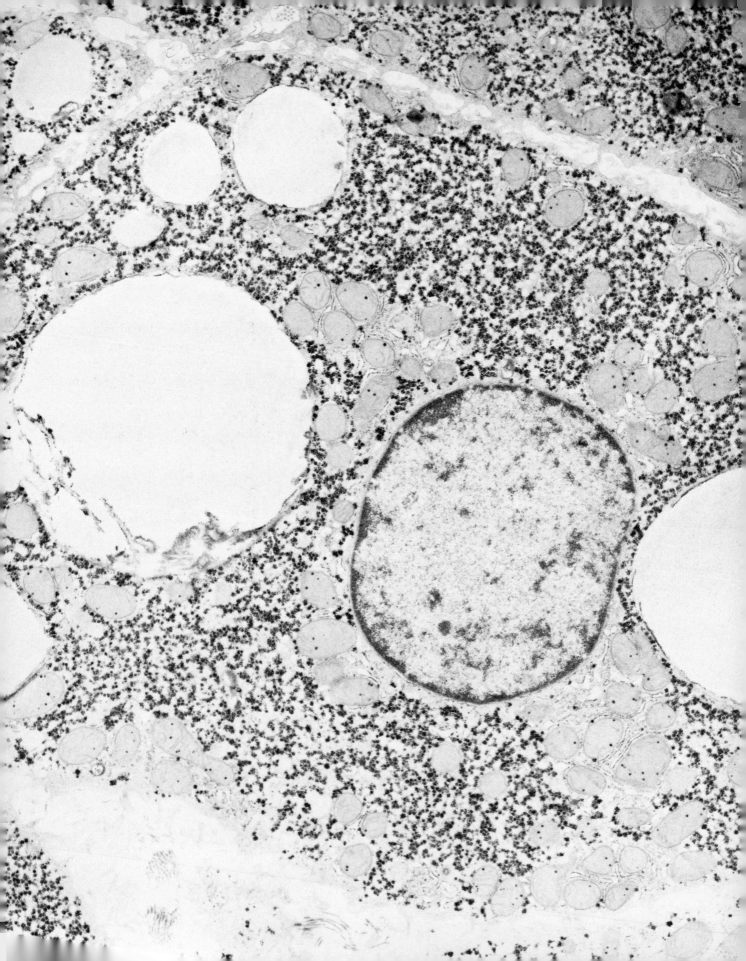

Pigmented Portal Macrophages of the Fatty Liver

FIGURE 52. The cell in this electron micrograph is a liver portal macrophage that contains two large masses of phagocytic material one above and the other below the compressed nucleus. The material consists of a characteristic combination of electron leucent globules within a very electron dense matrix. It appears to be a lipofucsin or ceroid type of material, that is, a residual body of phagocytic material. We believe that this material is derived from the fat phagocytized from necrotic cells in fatty livers. (11,000×)

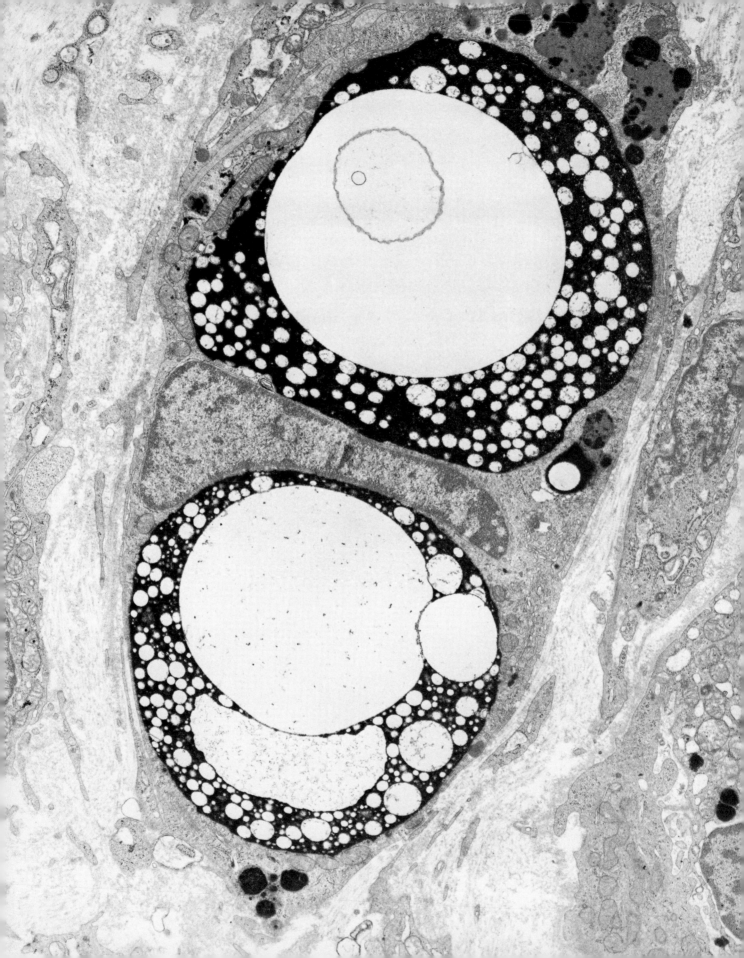

Cirrhosis

FIGURE 53. Cirrhosis is characterized by a loss of the normal hepatocyte plate to a sinusoid relationship. In a two-dimensional tissue section the normal relationship is single rows of hepatocytes separated by sinusoids. In cirrhosis the single row of hepatocytes is replaced by hepatocytes many cells thick. This relationship is readily appreciated at the low power ultrastructural level.

Another feature of cirrhosis is the microvillarization of the cell membranes and widening of the intercellular space between hepatocytes. (4900×)

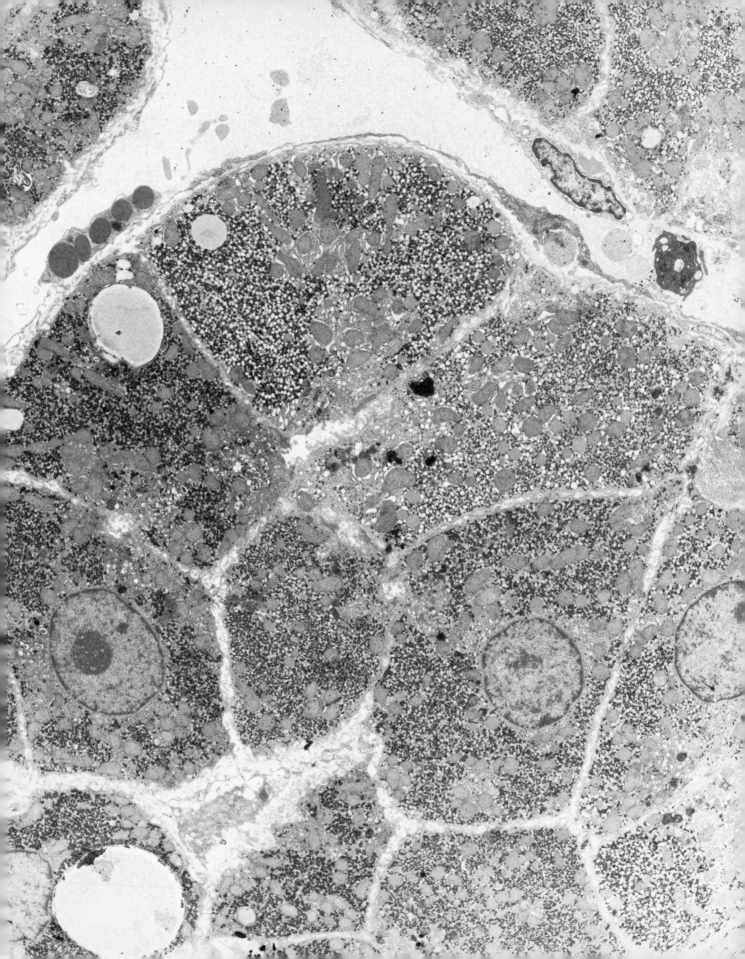

"Capillarization" of Hepatic Sinusoids and Fibrosis of the Space of Disse

FIGURE 54. The space of Disse is lined on one side by hepatocytes (open arrow) and on the other side by sinusoidal endothelial or Kupffer cells (closed arrow). None of these cells normally lie on a basement membrane except in the immediate vicinity of a portal area or a central vein. The occurrence of a basement membrane (at the tips of the arrows) in any other area is abnormal and is called capillarization of the sinusoids by Schaffner and Popper. (Gastroenterology 44:239, 1963.)

Another abnormality is the large accumulation of collagen which has widened the space of Disse. This fibrosis and the basement membrane are features of chronic liver disease and cirrhosis. (9300×)

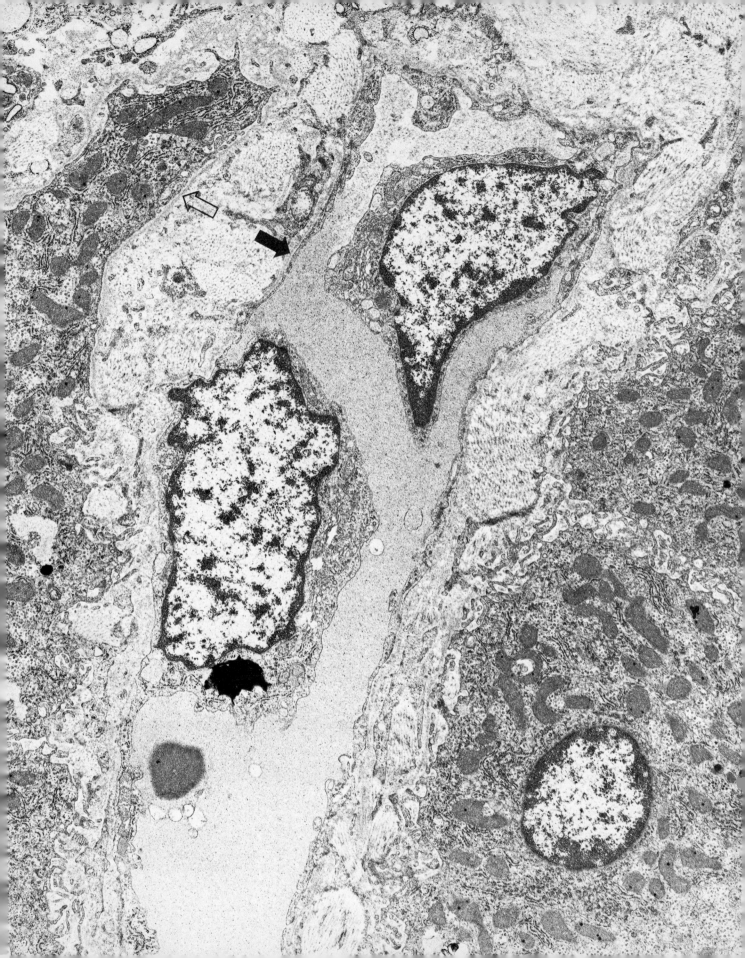

Liver Bile Plug

FIGURE 55.   The finely granular to flocculent material in this micrograph is encircled by plasma membranes of hepatocytes and is in a dilated bile canaliculus. The material is presumably bile. The microvilli of the normal canaliculus are nearly absent in this case as they are in most cases of cholestasis. (20,400×)

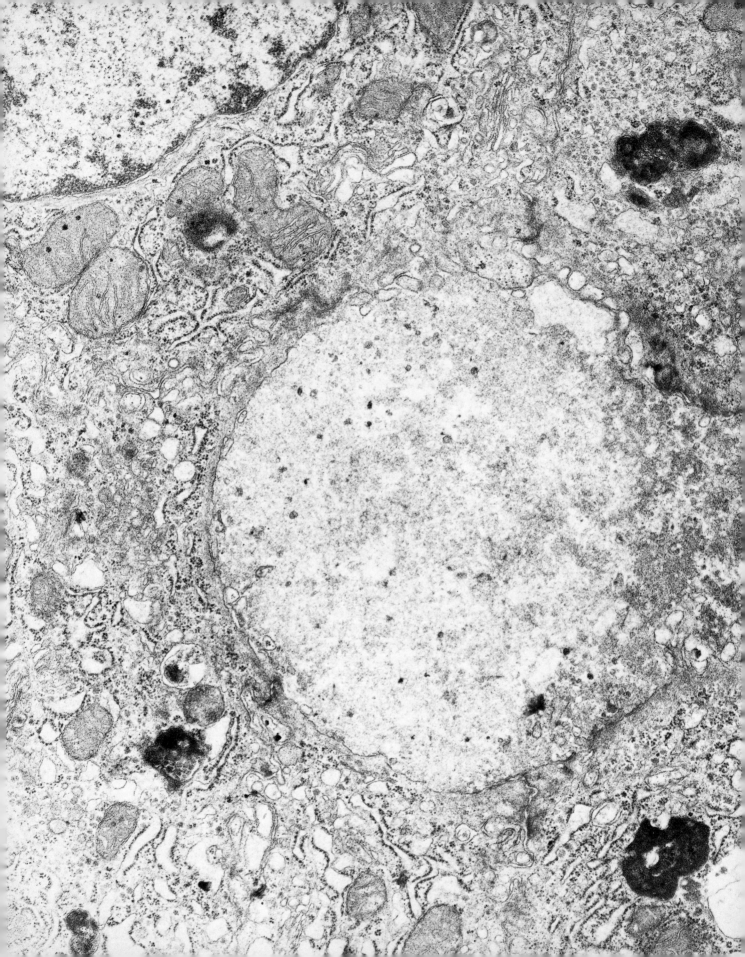

Liver Bile Plug

FIGURE 56. This dense accumulation of material is localized within the lumen of a bile canaliculus of the liver. Microvilli which normally extend into the lumen are not present; this deficiency is common in cholestasis. The material is the ultrastructural correlate of the bile identified by specific stains for light microscopy. This accumulation of bile is characterized by its finely granular to almost homogeneous appearance. There are also two to three tones of density within the bile plug. Bile presents itself ultrastructurally in many ways of which this is one example. Other examples are given elsewhere in this atlas. (32,800×)

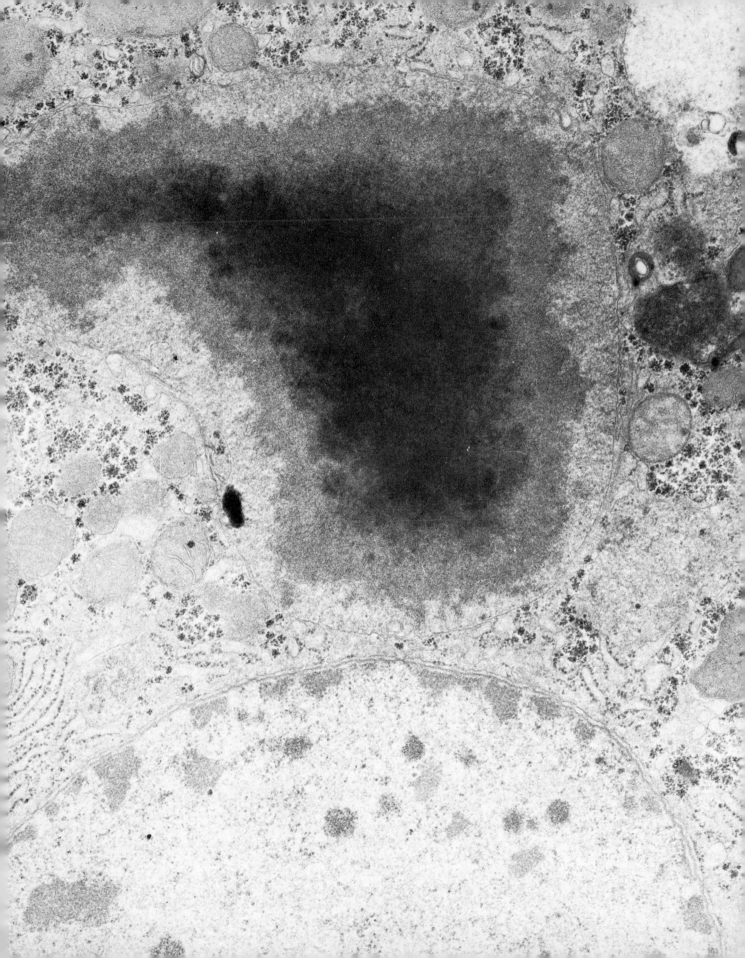

Intrahepatocytic Bile (Crystalloid Type)

FIGURE 57. The electron dense material demonstrated in this electron micrograph is free within the hepatocyte cytoplasm. The material is associated with marked cholestasis that is specifically demonstrated by light microscopy with bile stains. This bile is characterized by its heavy electron staining and membranous, myelin-like substructure. This is just one of many ways in which bile presents itself ultrastructurally. Other ways are given in subsequent electron micrographs. (33,000×)

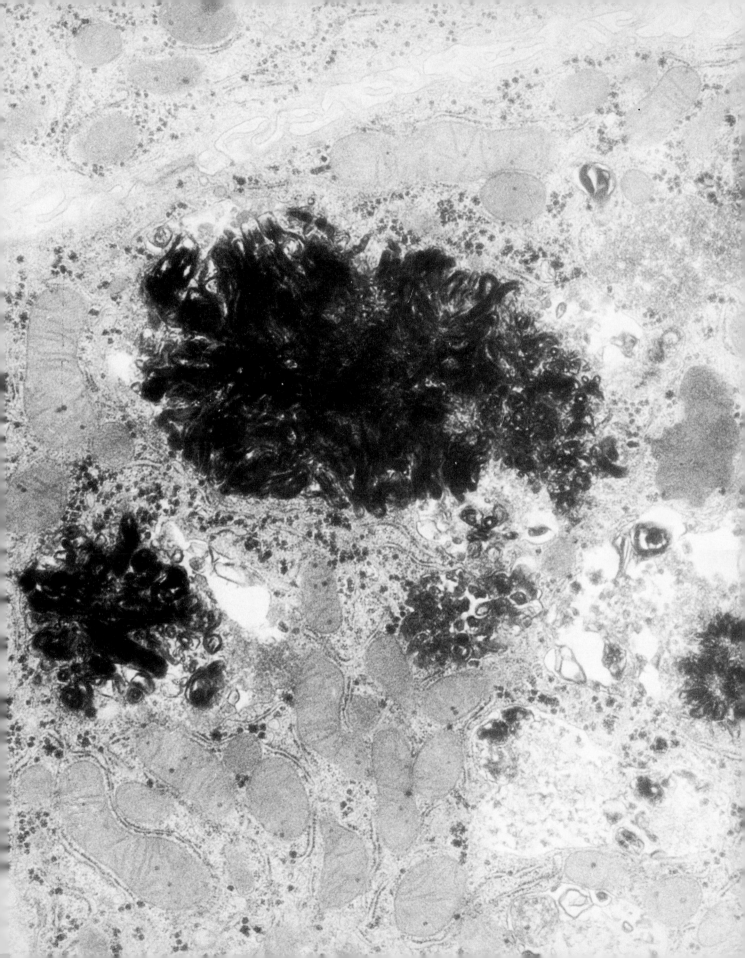

Intrahepatocytic Bile (Electron Lucent Type)

FIGURE 58.   Large deposits of relatively electron transparent material is present in the hepatocytes of this case of cholestasis. The deposits represent bile and are characterized by a wispy, membranous substructure as well as myelin-like membrane material. The hepatocytes in this electron micrograph have formed an acinus whose lumen contains many microvilli as demonstrated in the upper left corner. A glycogen body is located in the lower left corner. Glycogen bodies are often seen in cholestatic hepatocytes and are more clearly demonstrated in other micrographs. (19,700×)

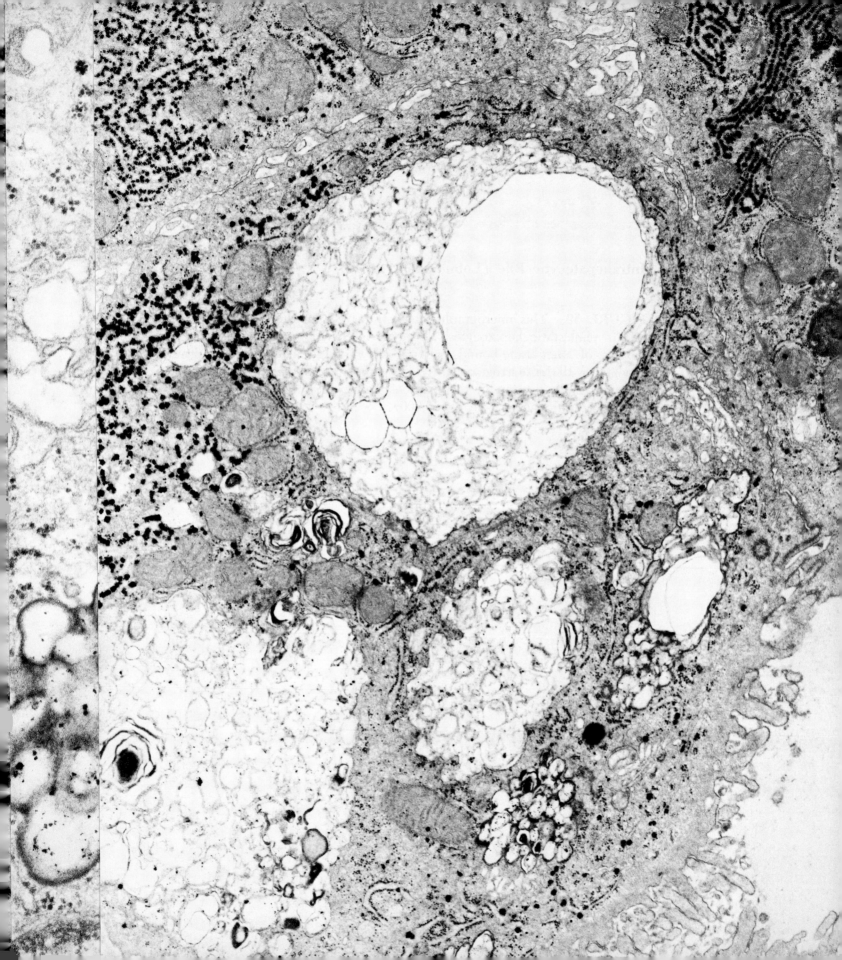

Hepatic Porphyrin Crystal

FIGURE 61. This electron micrograph demonstrates one of the many forms in which porphyrin has been reported in hepatocytes. This ultrastructural configuration has been compared with a chicken bone-like inclusion. (41,000×)

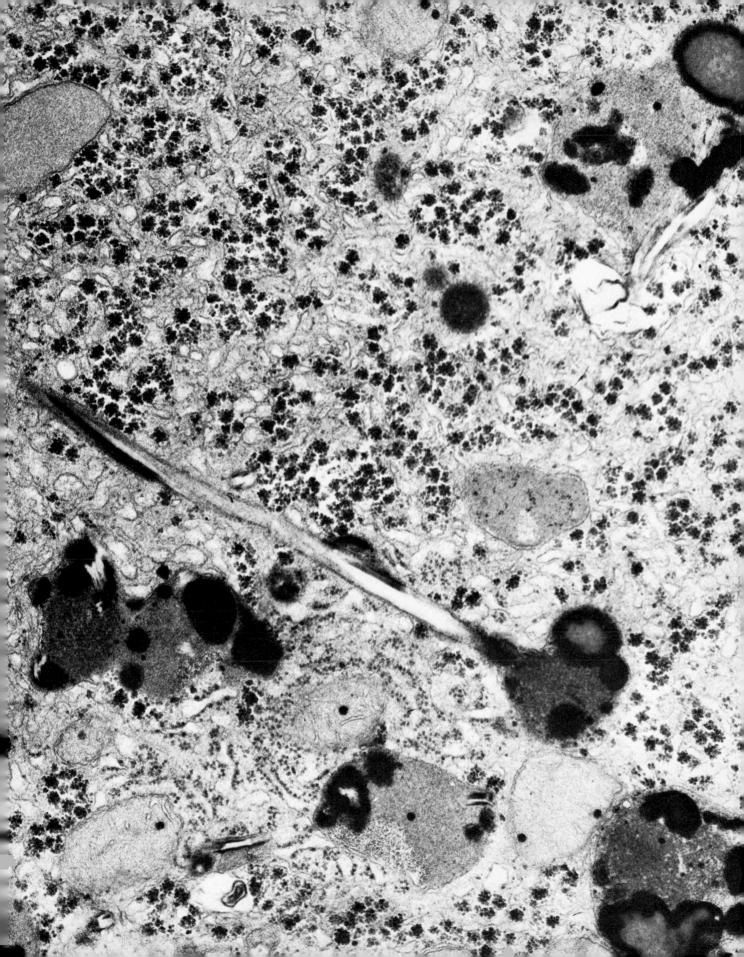

"Horse Tail" Inclusion in a Hepatocyte

FIGURE 62. Revealed here is an inclusion within the cytoplasm of a hepatocyte. The inclusion is slightly larger than a mitochrondrion. For want of a name for the inclusion, "horse tail" has been applied to it because of the similarity. The significance and nature of this type of inclusion are unknown. (55,000×)

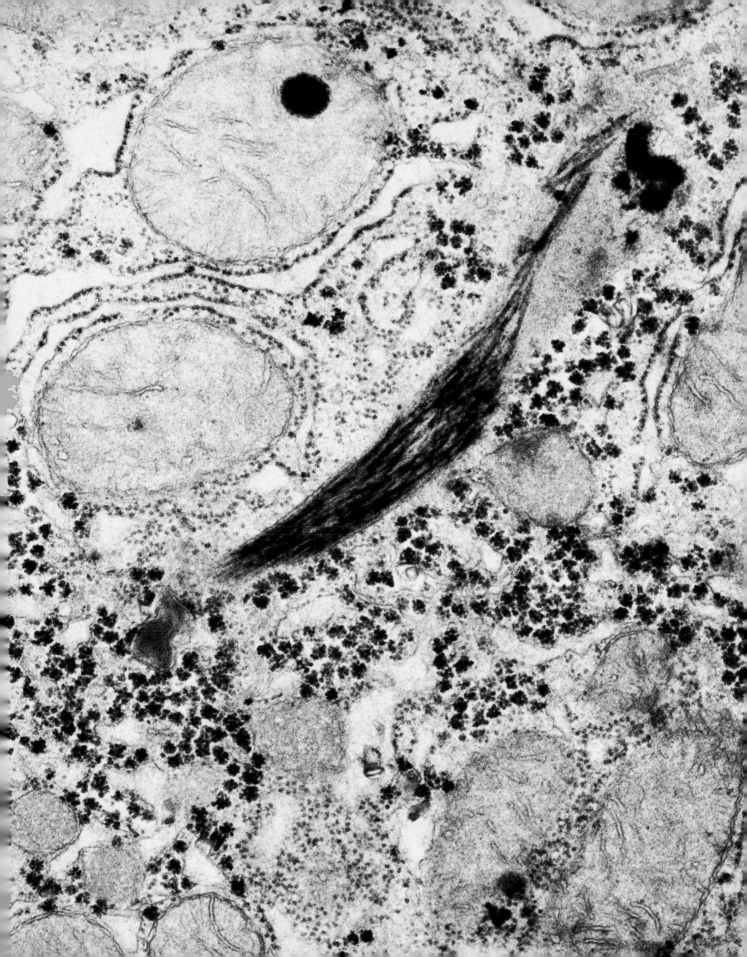

"Herring Bone" Crystal in a Hepatocyte

FIGURE 63. This very large inclusion in the cytoplasm of a hepatocyte has, upon close inspection, a "herring bone" pattern. The inclusion is not membrane bound, and it seems to occasionally form around normal cytoplasmic components like glycogen or a myelin figure as in this example. Neither the significance nor the nature of such inclusions is known, but they are found in various liver diseases. (40,000×)

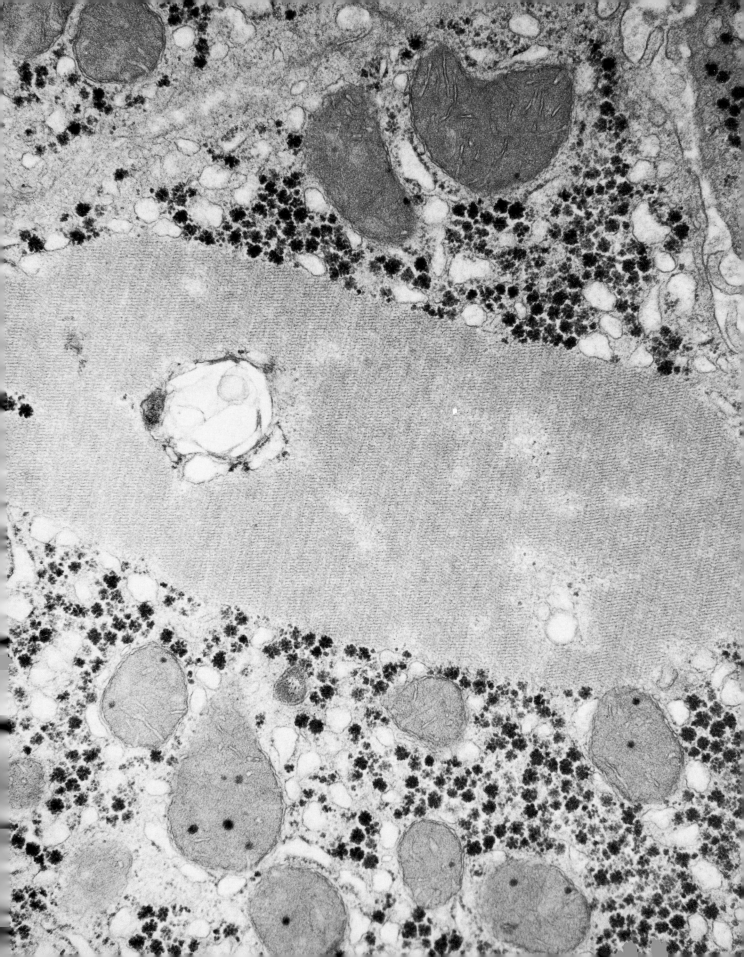

Talc Crystals in Liver Macrophages

FIGURE 64. Numerous electron dense, elongated particles are present within the liver macrophages of the micrograph. The particles are from talc that is inadvertently injected along with drugs by narcotic users. The talc is most often associated with section tears and with the initiation of knife marks (arrow) presumably caused by talc's mineral structure. (4700×)

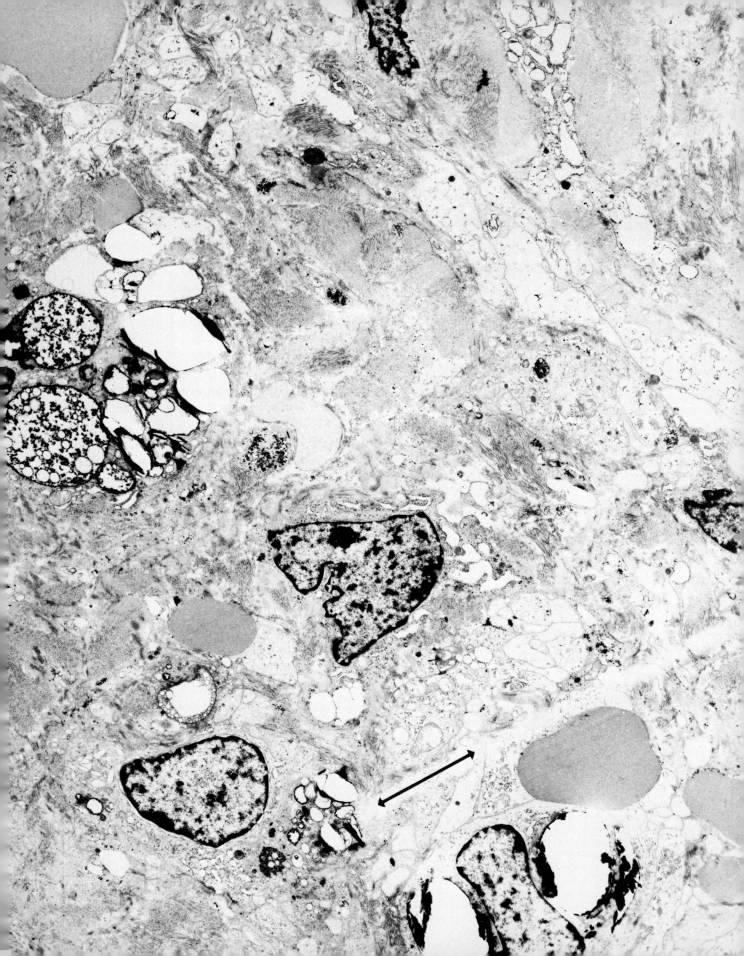

Hepatocyte Glycogen Body and Condensed Mitochondria (Low Magnification)

FIGURE 65. The hepatocyte in this figure demonstrates two abnormal features: a glycogen body and condensed mitochondria.

A glycogen body consists of alternating monolayers of glycogen and lamellar smooth endoplasmic reticulum (SER). The structural association of glycogen and SER reflects the functional association of glycogen metabolism with the SER. Glycogen bodies are often but not exclusively found in hepatocytes of patients with cholestasis.

Condensed mitochondria are characterized by a dense matrix and widened intracristal spaces. Condensation is associated with an increased ratio of ADP/ATP, which can be caused by a large demand for ATP or a defect in its regeneration from ADP. The latter defect can result from an acute cellular injury. A higher magnification of a similar area is provided in the next figure. (14,000×)

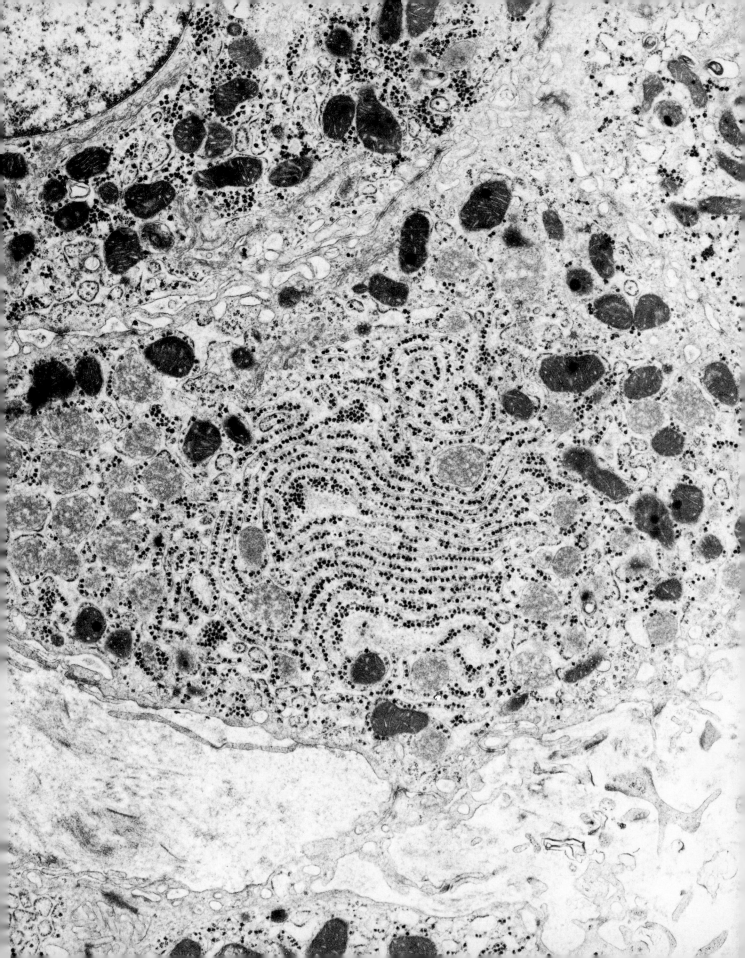

**Hepatocyte Glycogen Body and Condensed
Mitochondria (High Magnification)**

FIGURE 66.   A portion of a glycogen body is presented here. The lamellar
smooth membrane and the glycogen are clearly resolved. The mitochondria
are condensed. Their expanded cristae and dense matrix are prominent.
(29,100×)

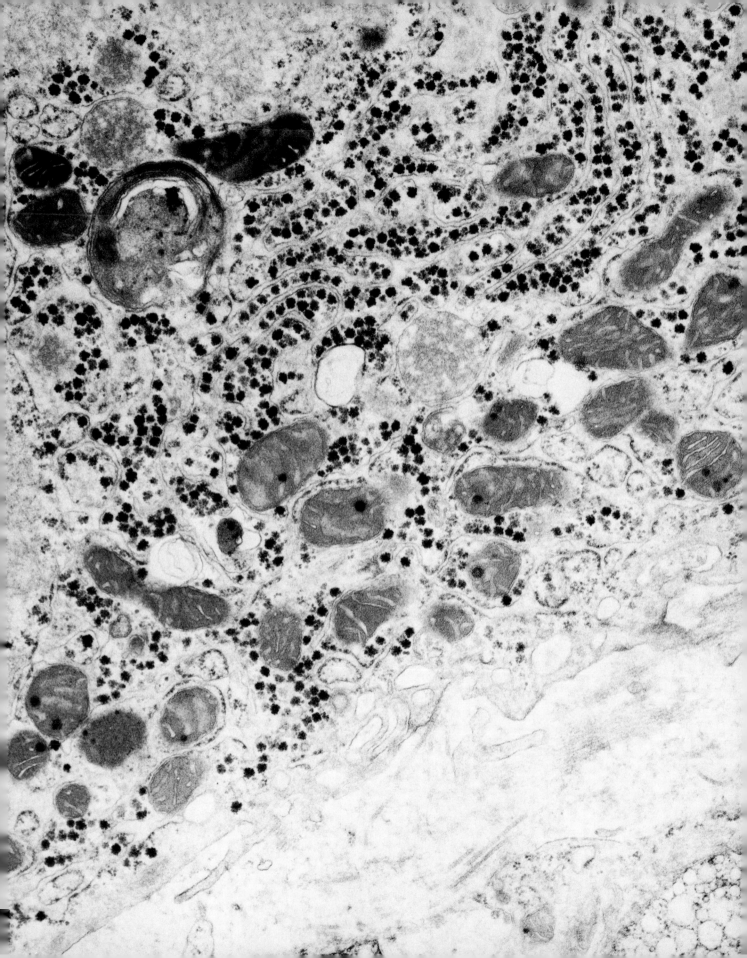

Von Gierke's Disease (Type I Glycogenosis)

FIGURE 67.   This liver biopsy specimen is from a 7-month-old baby with relatively asymptomatic hepatic enlargement. The liver cell is distended with alpha glycogen particles (most of the glycogen appears partially autolyted), and other organelles are displaced to the periphery. A few fat droplets are seen in the cytoplasmic sap. Although Type I glycogenosis is the most common type with few symptoms at the onset, the ultimate diagnosis depends on histochemical or biochemical determination of hepatic enzymes. The fine structure itself is not conclusive. (38,300×)

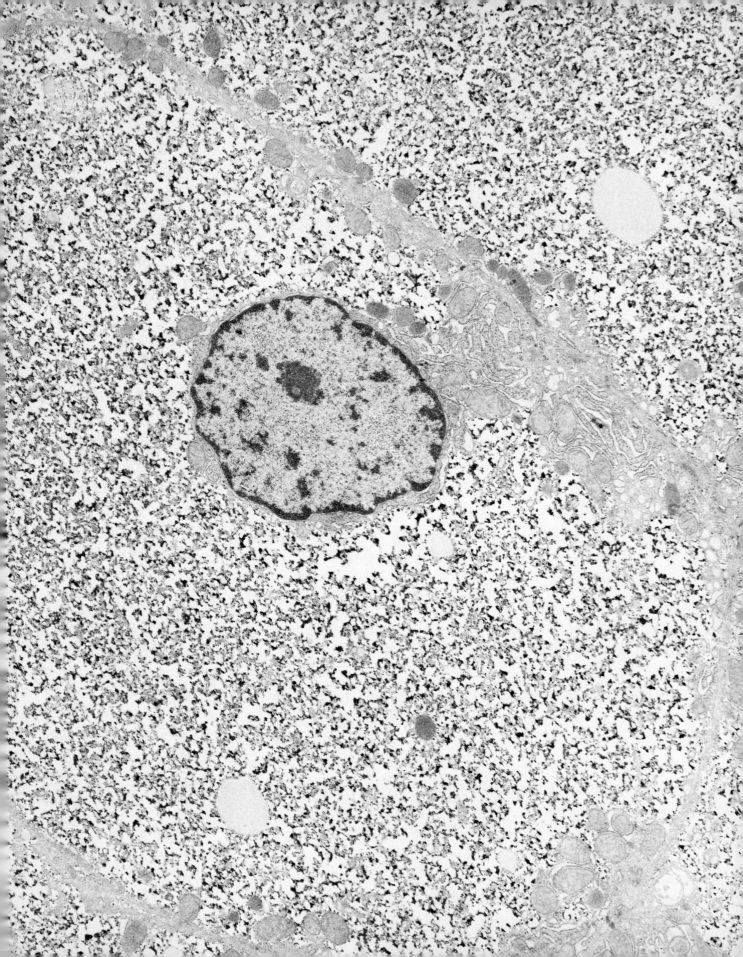

Cholesteryl Ester Storage Disease

FIGURE 68. Portions of liver cells from a 15-year-old white male with an asymptomatic hepatic enlargement are shown. Note a collection of lipid droplets with well-defined membranes. Other cell organelles appear normal. The disease is characterized by an accumulation of abundant cholesterol esters and triglycerides in the liver. Therefore, an examination of an unstained frozen-section for polarizing material or examination by histochemical study is extremely useful. Grossly, the liver is orange in color. The disease is associated with a deficiency in acid lipases in leukocytes. (22,000×)

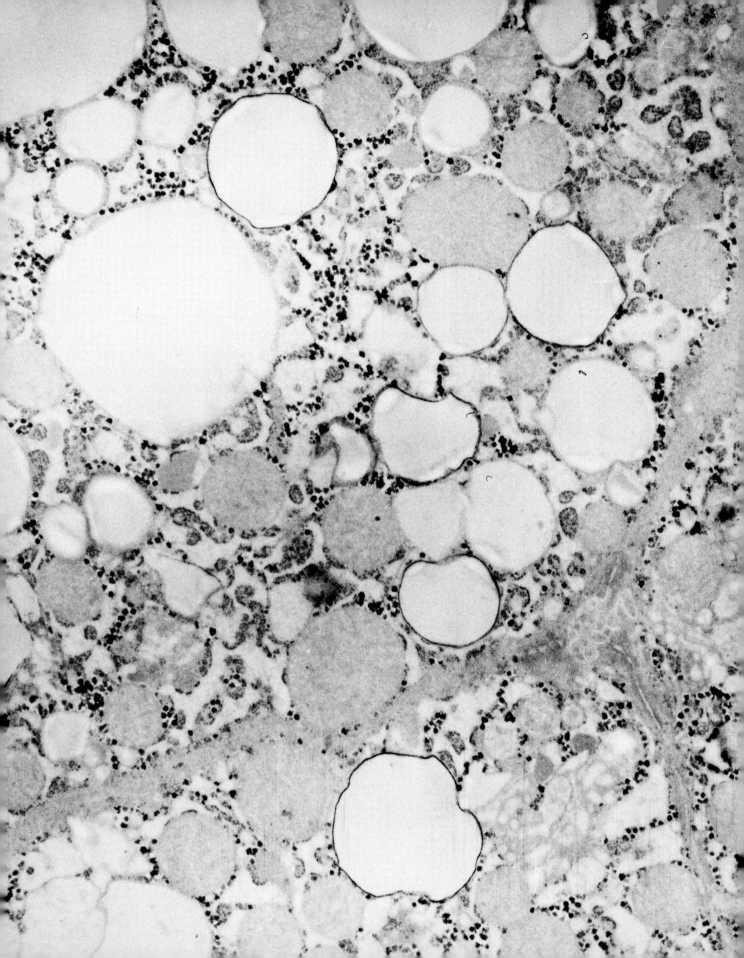

RER Lake

FIGURE 69.   In this electron micrograph there is an abnormally dilated cisterna of endoplasmic reticulum along with other less dilated but still abnormal ones. The endoplasmic reticulum is studded with ribosomes and therefore is rough endoplasmic reticulum (RER). A common term for such dilated RER is a "RER lake." The cisternae are filled with a finely granular material which is presumably protein synthesized by the RER. The nature of the protein and the reason for its abnormal accumulation are unknown.

This example of a RER lake is from a hepatocellular carcinoma. Other examples of its occurrence are hepatocytes in alpha-1-antitrypsin deficiency and plasma cells with Russell bodies. (18,700×)

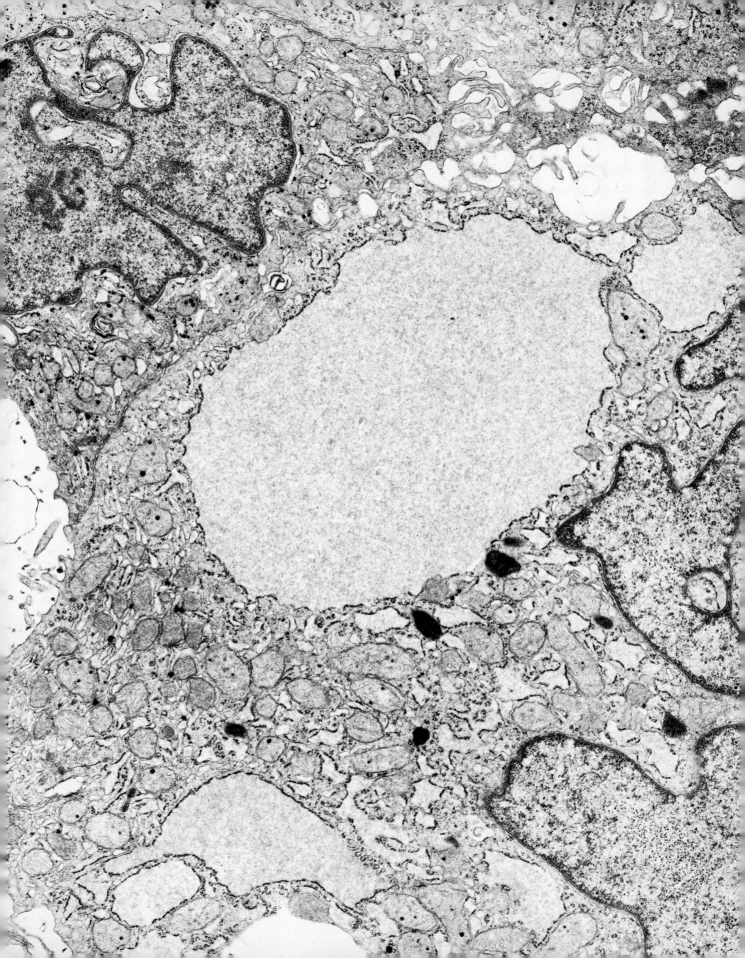

The Liver in Alpha-1-Antitrypsin Deficiency

FIGURE 70. This is a portion of a hepatocyte from an infant with conjugated hyperbilirubinemia. The endoplasmic reticulum is distended with unusual amorphous material which has been shown to be alpha-1-antitrypsin immunochemically. This condition is one of the common causes of neonatal conjugated hyperbilirubinemia and, at times, difficult clinically to differentiate from neonatal hepatitis. As alpha-1-antitrypsin constitutes a major portion of serum alpha-1-globulin, serum protein electrophoresis is useful in ruling out this condition. (24,500×)

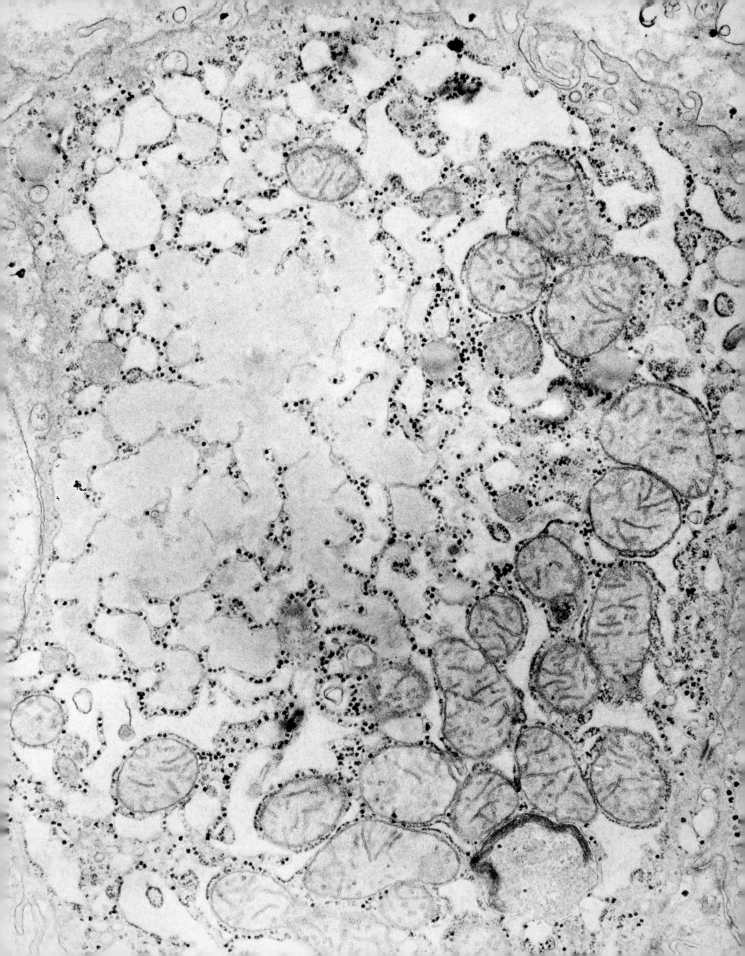

Fat-Storing Cell—Hypervitaminosis A

FIGURE 71. Although the fat storing cell is a normal component of the liver, it is increased in number and size in various pathological conditions of the liver. Most remarkable is the increase found in hypervitaminosis A. The Ito cell is seen in the lower left corner of this figure. The cell is usually found between hepatocytes at the space of Disse. The fat droplets that characterize the normal fat storing cells are easily distinguished from abnormal hepatocyte fat droplets by comparing their location and their electron density. (10,400×)

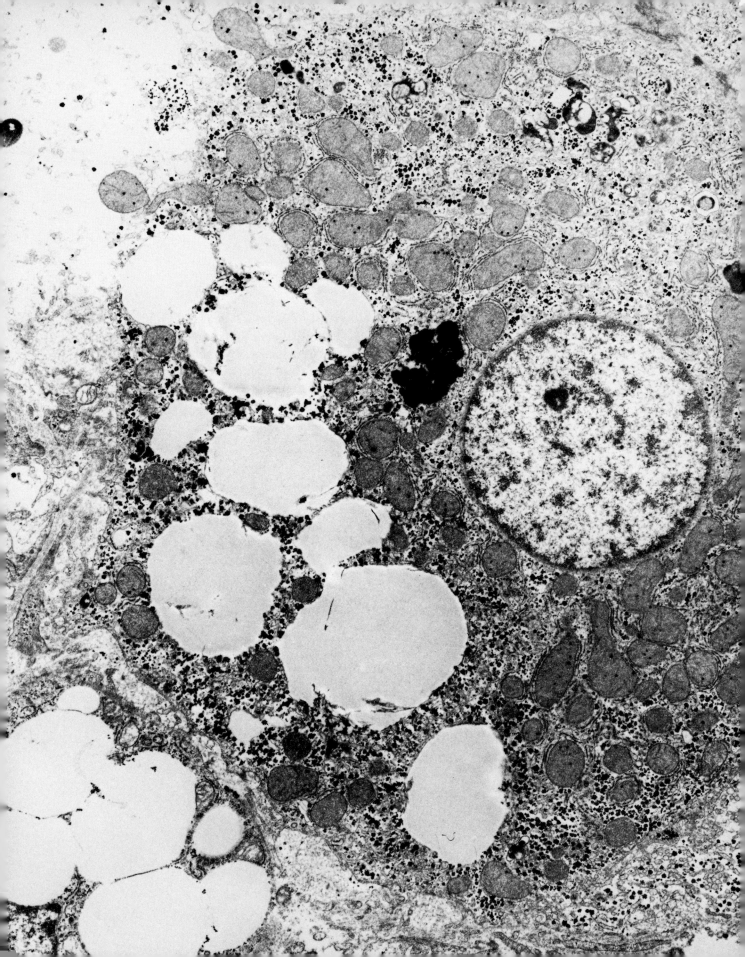

The Liver in Reye's Syndrome

FIGURE 72.   A portion of a hepatocyte from a 6-year-old child with acute onset of vomiting and lethargy. There is generalized swelling of mitochondria (M) with panlobular fatty metamorphosis. Note the mitochondria are distended, irregular in configuration and more importantly with rarefaction of the matrix. The glycogen is slightly depleted. The remaining cytoplasmic organelles are relatively well preserved. (30,000×)

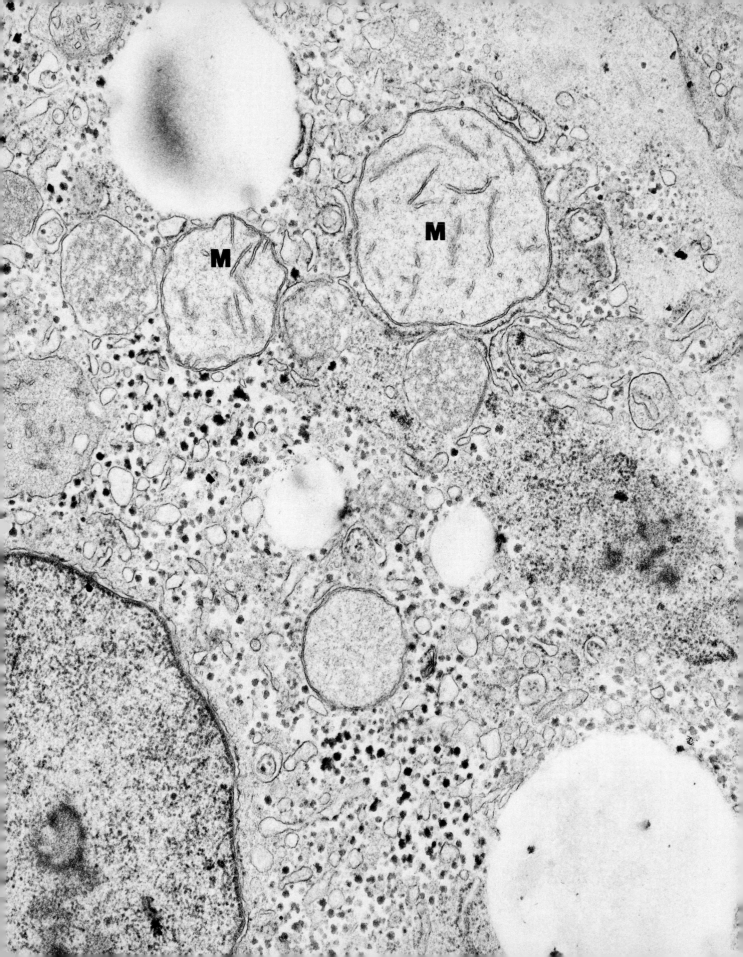

Liver

Hepatocellular Carcinoma

FIGURE 73. Certain neoplasms of the liver show ultrastructural features that characterize them as probable hepatocellular carcinomas. In the neoplasm presented here the alpha glycogen and the round mitochondria with dense matrix and few cristae are features normally found in hepatocytes thus strongly indicating a well-differentiated primary hepatocellular carcinoma. (17,000×)

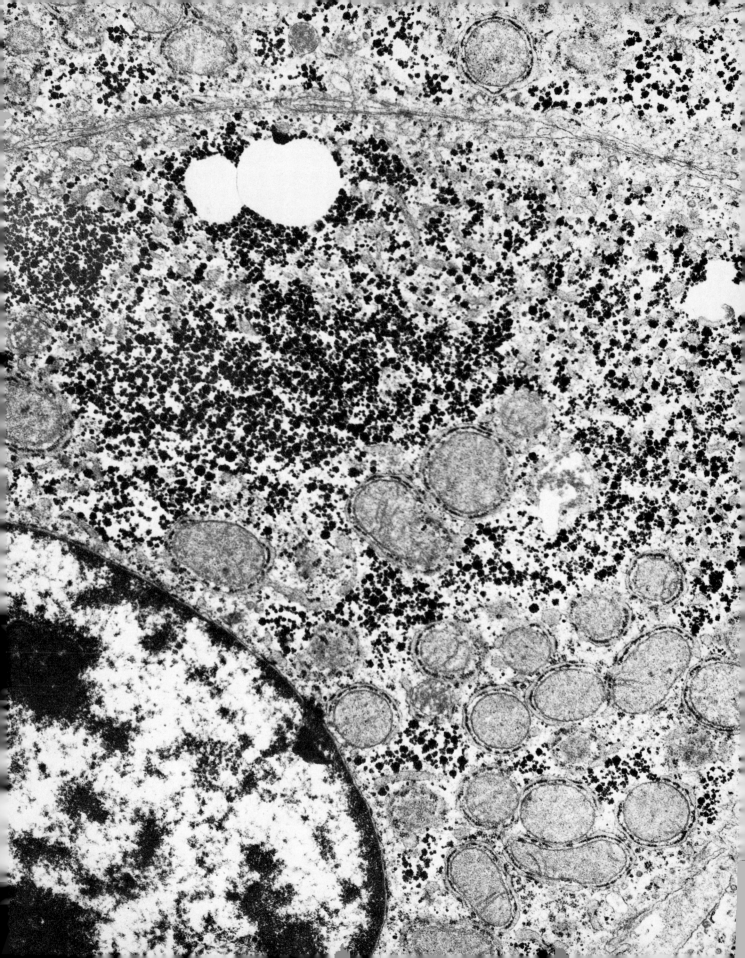

Hepatocellular Carcinoma

FIGURE 74. This micrograph presents a carcinoma cell with round nucleus and nucleolus. The cytoplasm contains round mitochondria with a few cristae, dilated endoplasmic reticulum, and scattered glycogen particles. (19,200×)

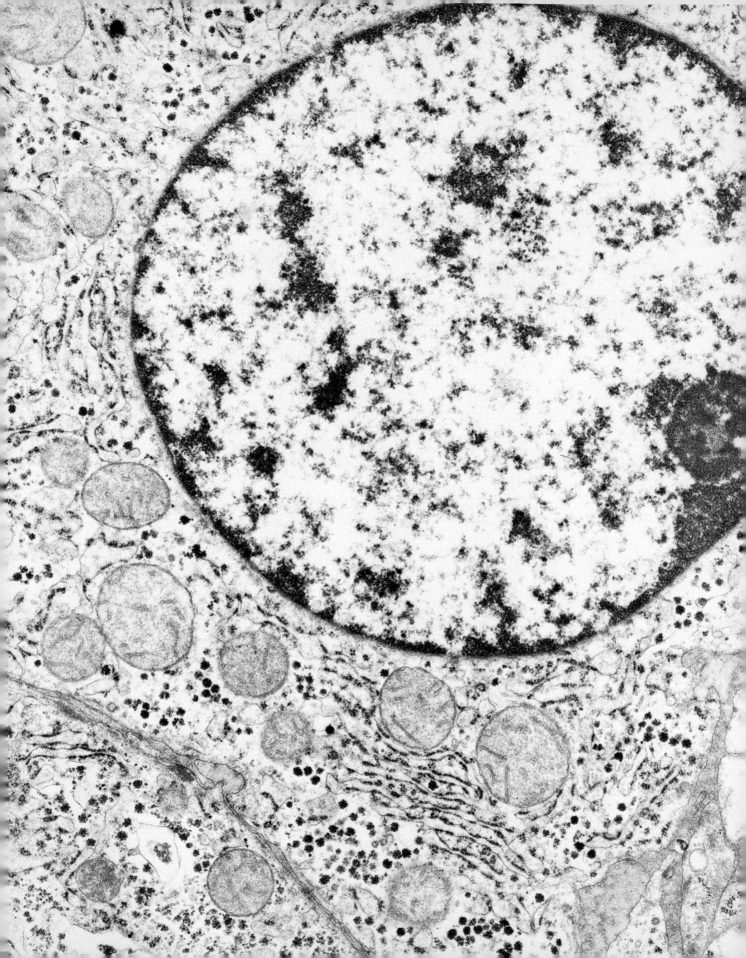

Hepatocellular Carcinoma

FIGURE 75. This carcinoma cell has a large nucleus with surface indentation, clumped marginal chromatin and a very prominent nucleolus. The cytoplasm is rich in round swollen mitochondria, dilated endoplasmic reticulum and scattered glycogen granules. (22,500×)

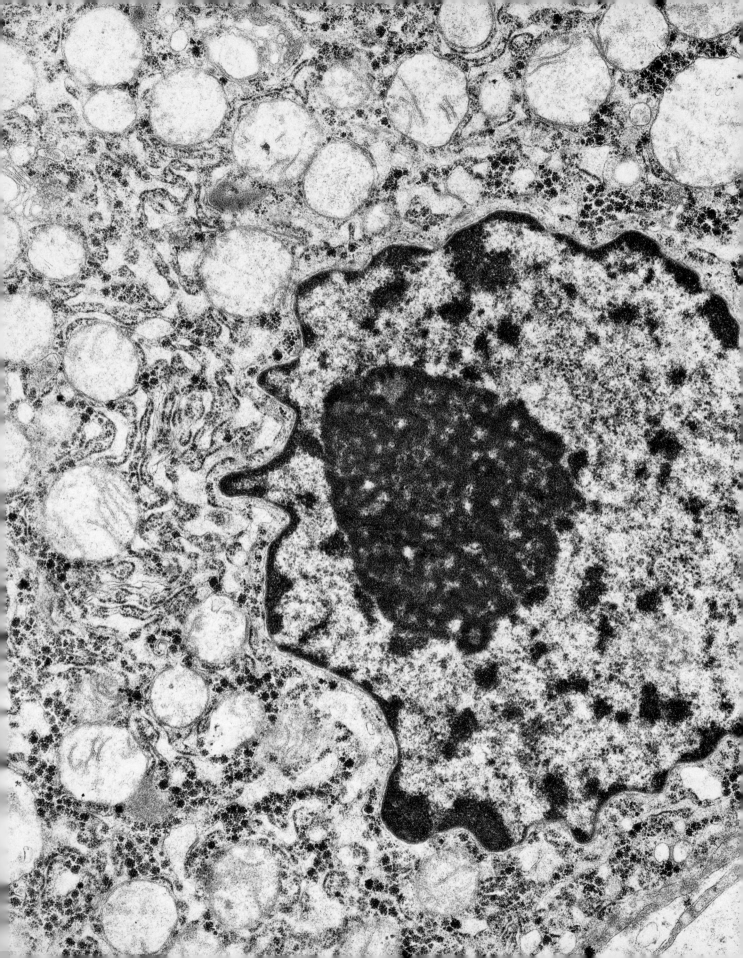

Hepatocellular Carcinoma (With Mallory Body)

FIGURE 76. A mass of randomly arranged filaments is present throughout this portion of cytoplasm of a neoplastic cell. The filaments are approximately 15 nm in diameter. The mass is identical to Mallory bodies often found in pathological liver hepatocytes (see examples in other figures in this atlas). The presence of Mallory bodies in neoplastic cells is not often encountered, but when present it does mark the neoplasm as a hepatocellular carcinoma. (48,300×)

Hepatoblastoma

FIGURE 77. This is a malignant tumor of the liver from a 3-year-old boy who did well following lobectomy. The neoplastic cells are arranged in sheets, trabeculae, or organoid patterns and form occasional poorly formed bile canaliculi and sinusoids. Like normal hepatocytes, a Golgi apparatus is seen near the bile canaliculus. Cytoplasmic organelles are relatively sparse and principally constitute endoplasmic reticulum, ribosomes, and mitochondria. No obvious microbodies are seen. The nuclei are relatively large, irregular in shape and contain multiple prominent nucleoli. Hepatoblastoma is the most common malignant tumor of the liver in early childhood. If the tumor is localized in one lobe and can be surgically resected, the prognosis is favorable. The morphology and biology of this tumor is quite different from more aggressive hepatoma, seen mostly in adults. ($16,800\times$)

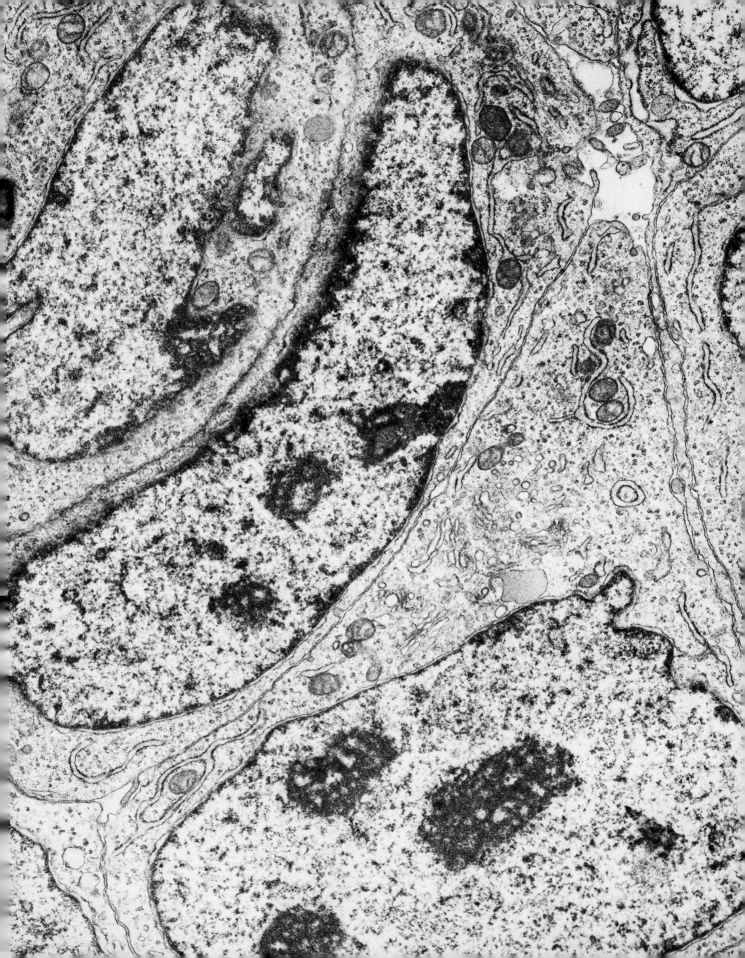

Hepatoblastoma

FIGURE 78. This is a high magnification of portions of neoplastic cells. Note that several cells are linked to each other by fairly well-formed tight junctions and form bile canaliculi. Note that the normal looking microvilli are seen along the bile canaliculi. (22,800×)

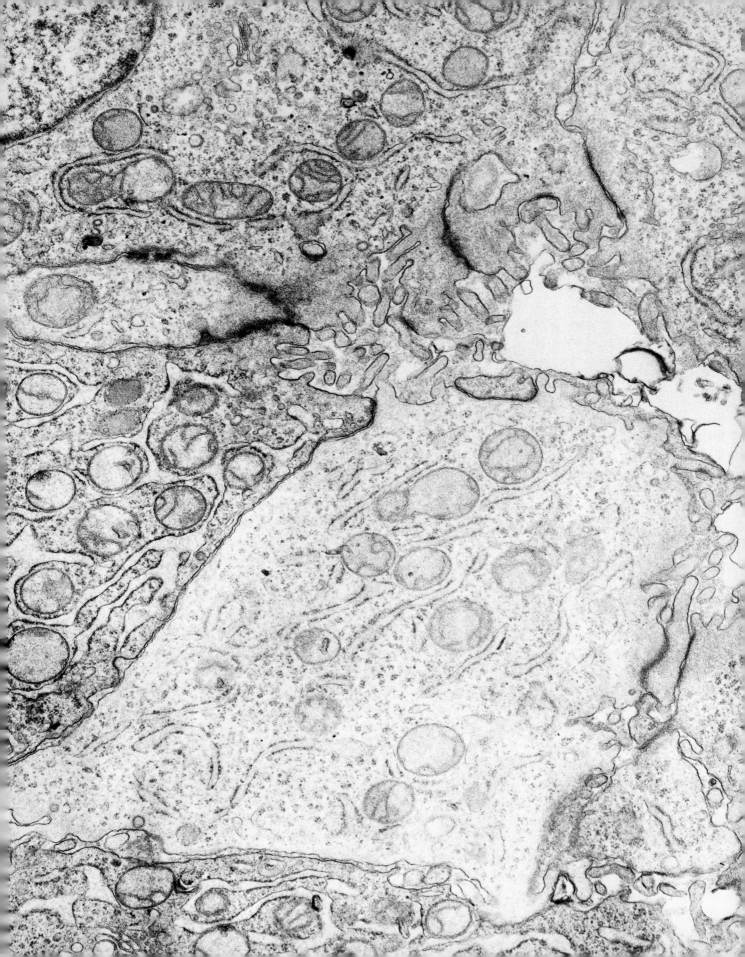

6 Gall Bladder

Adenocarcinoma of the Gall Bladder

FIGURE 79. The tumor cells are columnar to cuboidal cells with occasional microvilli on the luminal surface. Some of the cells are devoid of microvilli. In the apical portion of the cells there are many membrane-bound electron dense granules. One of them is protruding into the lumen. Other organelles of these tumor cells are mitochondria and rough endoplasmic reticulum which are often markedly dilated (not shown in this picture). (16,900×)

158

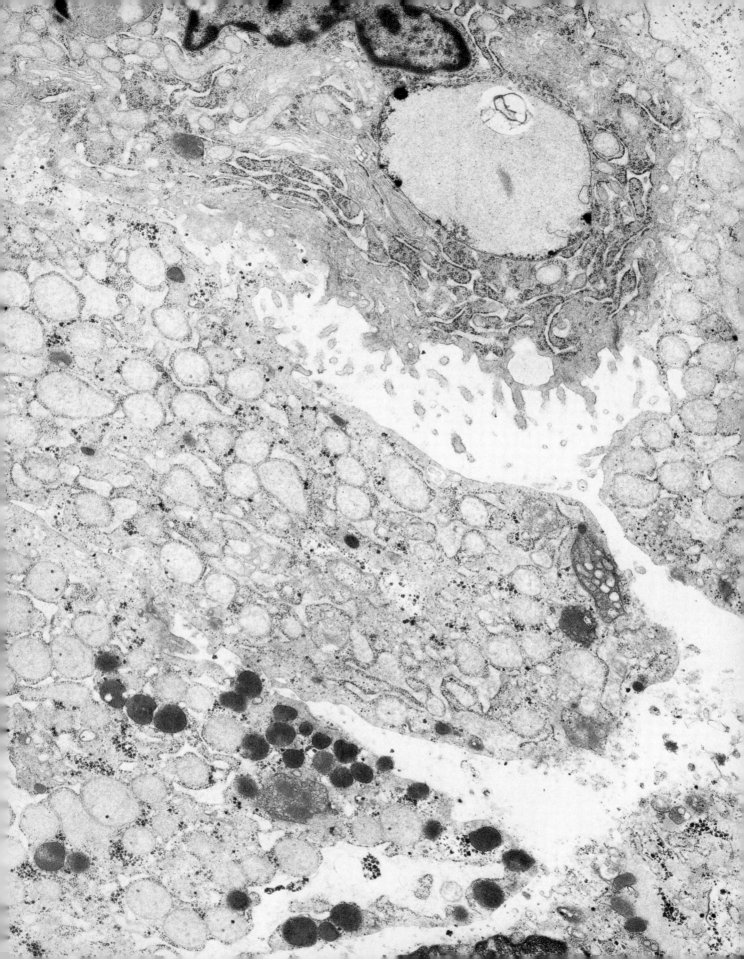

Adenocarcinoma of the Gall Bladder

FIGURE 80. A tumor cell shows many microvilli which are studded with electron dense material. The membrane bound granules in the apical cytoplasm are clearly demonstrated in this picture and show a density similar to that seen in the material over the microvilli. The material is mucicarmine positive under the light microscope. (24,600×)

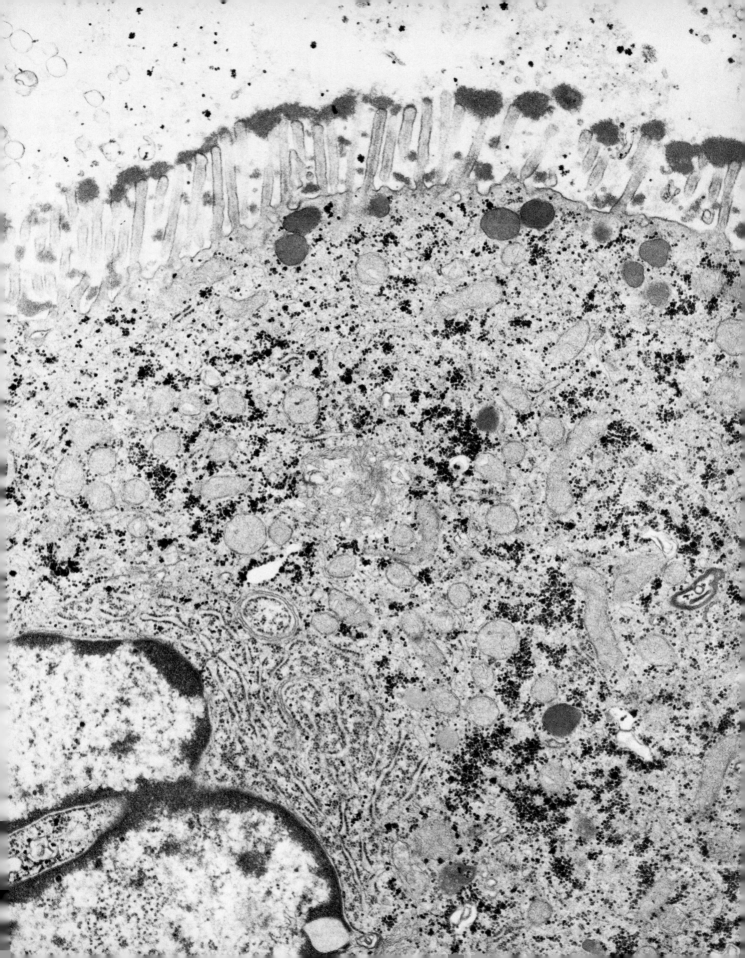

7 Pancreas

Beta-Cell Tumor (Insulinoma)

FIGURE 81. Several neoplastic cells from a beta-cell tumor show variable amounts of electron lucent cytoplasm containing various types of organelles. The most outstanding characteristic of these organelles consists of numerous dense core secretory granules of different sizes and shapes, distributed randomly throughout the cytoplasm. (19,200×)

162

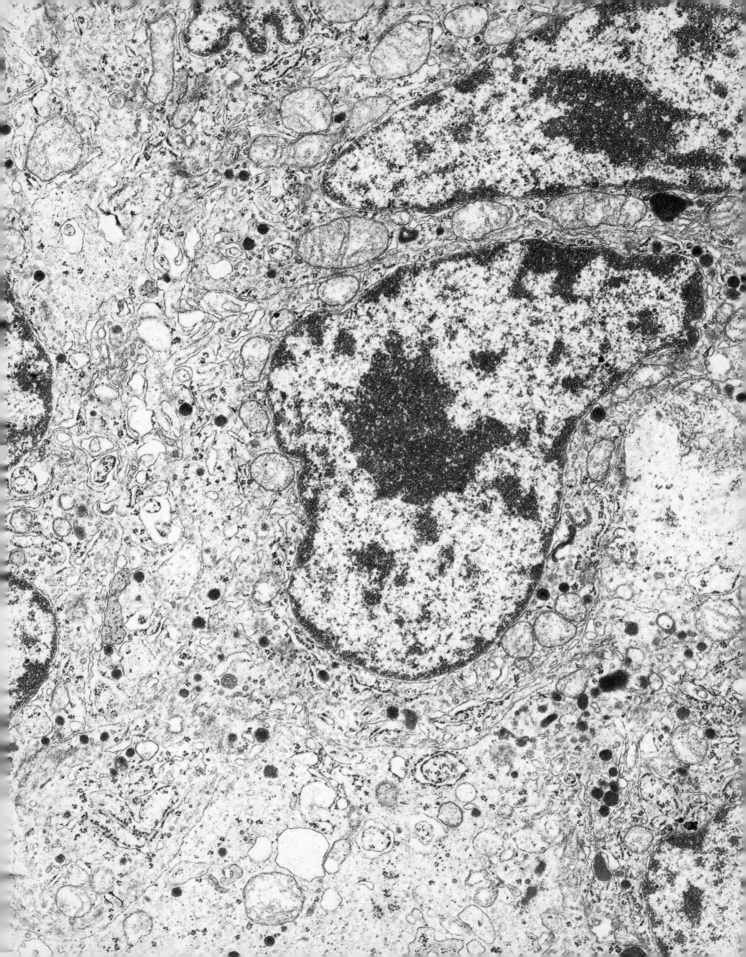

Beta-Cell Tumor (High Magnification)

FIGURE 82.   High magnification shows a membrane closely investing the homogeneous granule core. Most of the granules are round and their sizes vary from 120 to 140 nm. Several of them are triangular, while others display an irregular angular contour. They probably represent immature granules. The typical beta granules are somewhat larger than the alpha granules, and show a crystalline appearance and a wide halo surrounded by a unit membrane. The larger heterogeneous electron dense bodies represent lysosomes. (40,600×)

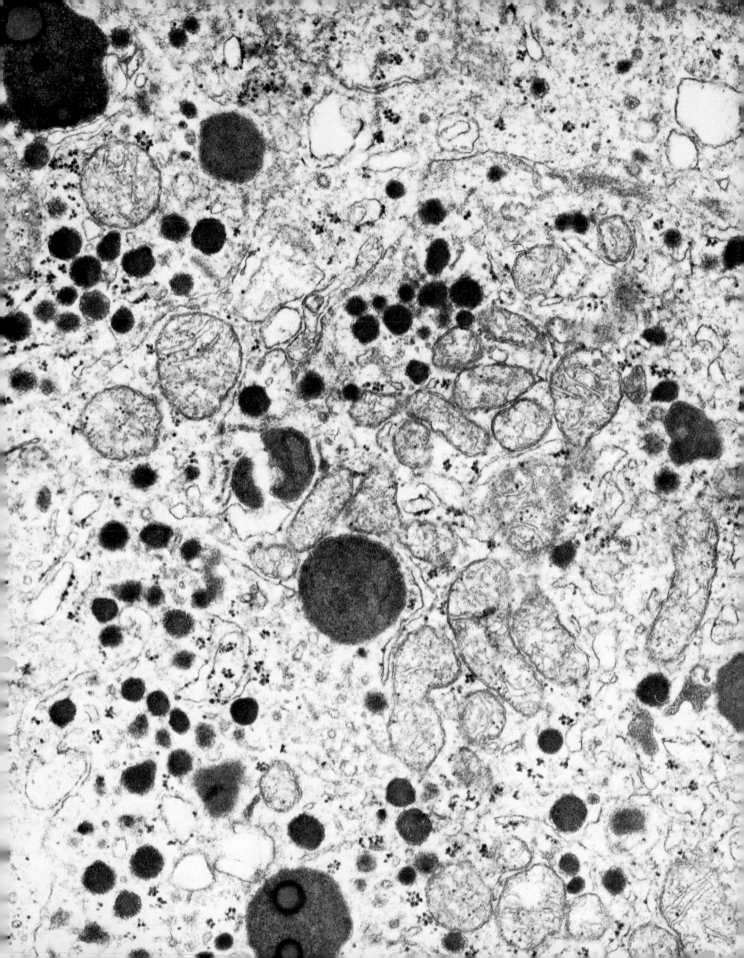

Alpha-Cell Tumor (Glucagonoma)

FIGURE 83. Cells from an alpha-cell tumor have slightly pleomorphic nuclei, inconspicuous rough endoplasmic reticulum, and numerous secretory granules which tend to be focally concentrated at times. Often the granules are concentrated at the vascular pole of the cells.

The typical granules have a dense osmiophilic spherical core with a less dense periphery closely surrounded by a single limiting membrane. They resemble the granule of normal alpha cells. Although variable in size, the largest profile measures about 250 nm. They are usually smaller and more uniform than the beta granules.

Similar secretory granules are found in many polypeptide synthesizing neoplasms which belong to the APUD (Amine Precursor Uptake and Decarboxylation) system, such as medullary carcinoma of thyroid, paraganglioma (chemodectoma), carcinoid, and oat cell carcinoma. Therefore, the diagnosis should take into account the clinical data, specific radioimmunoassays and immunofluorescence in order to identify the peptide hormone (or tumor markers) elaborated by the neoplasm. (9200×)

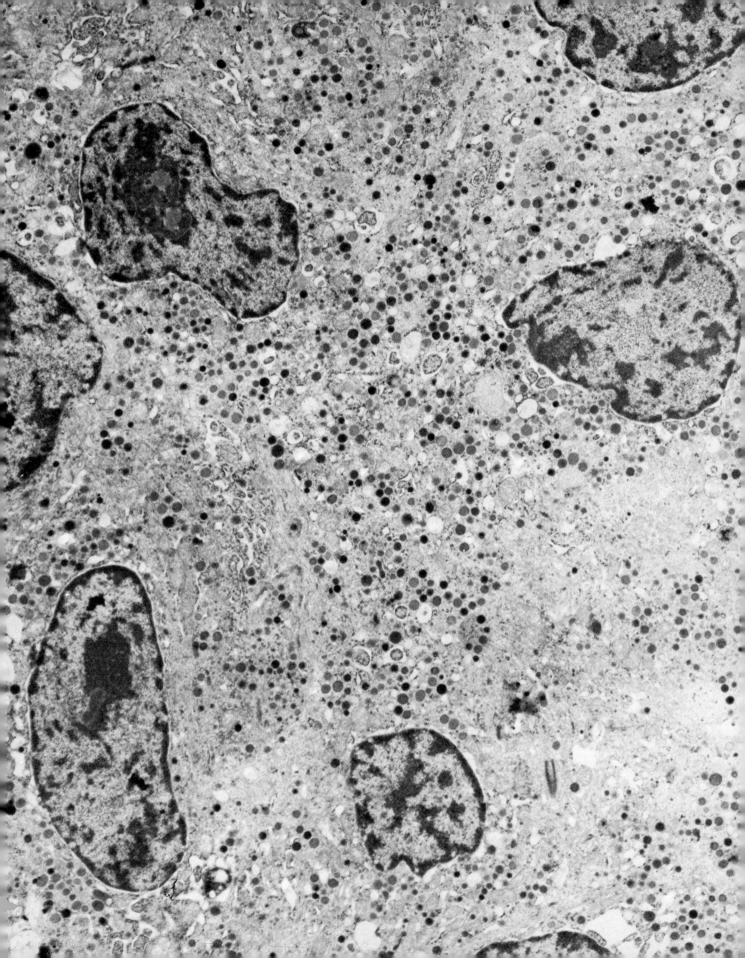

8 Pituitary Gland

Acidophilic Adenoma of the Pituitary Gland

FIGURE 84. Portions of three acidophilic cells from a child with acromegaly are presented. The tumor cells are degranulated with only a few scattered residual secretory granules in the periphery of the cells. The granules are larger than those seen in ACTH-producing tumors and are consistent with growth hormone-producing granules. The cells contain moderate numbers of mitochondria, rough endoplasmic reticulum, and small Golgi apparatus. It has been shown that the tumor cells tend to be degranulated when the hormone in the circulation is high. The size and configuration of secretory granules are often useful in differentiating the type of hormone produced; however, biochemical or immunoelectronmicroscopy study is ultimately required. (17,800×)

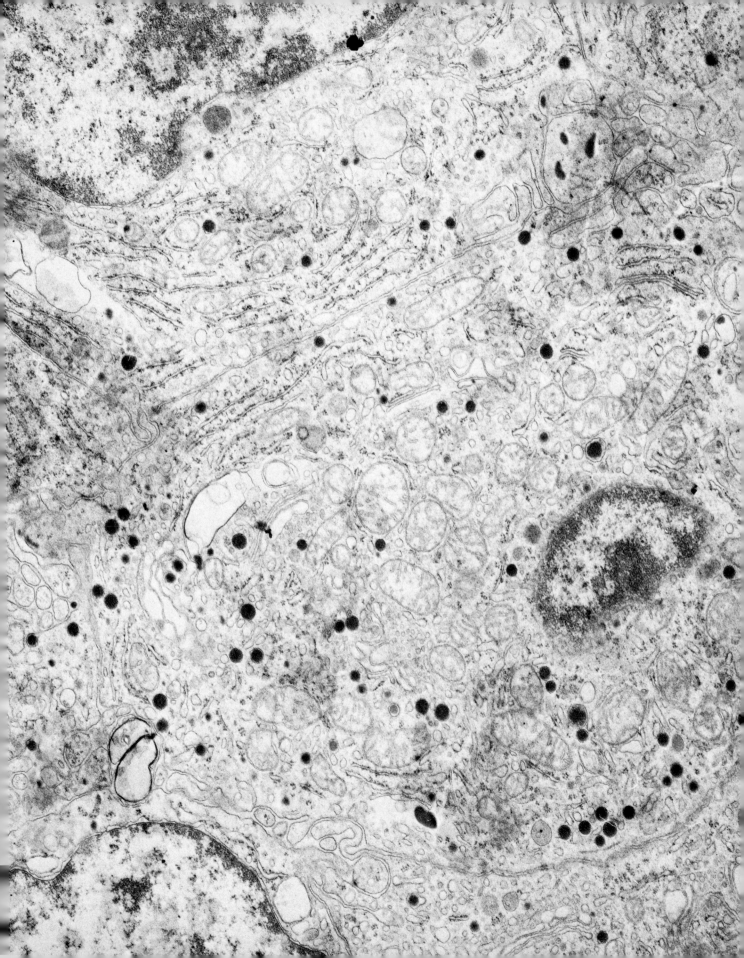

Acidophilic Adenoma of the Pituitary Gland

FIGURE 85. This is a high magnification of the peripheral portion of the cell shown in the preceding picture. Note the secretory granules have a dense core surrounded by a halo. The granules measure up to 200 nm in diameter which is consistent with growth hormone producing granules. (18,000×)

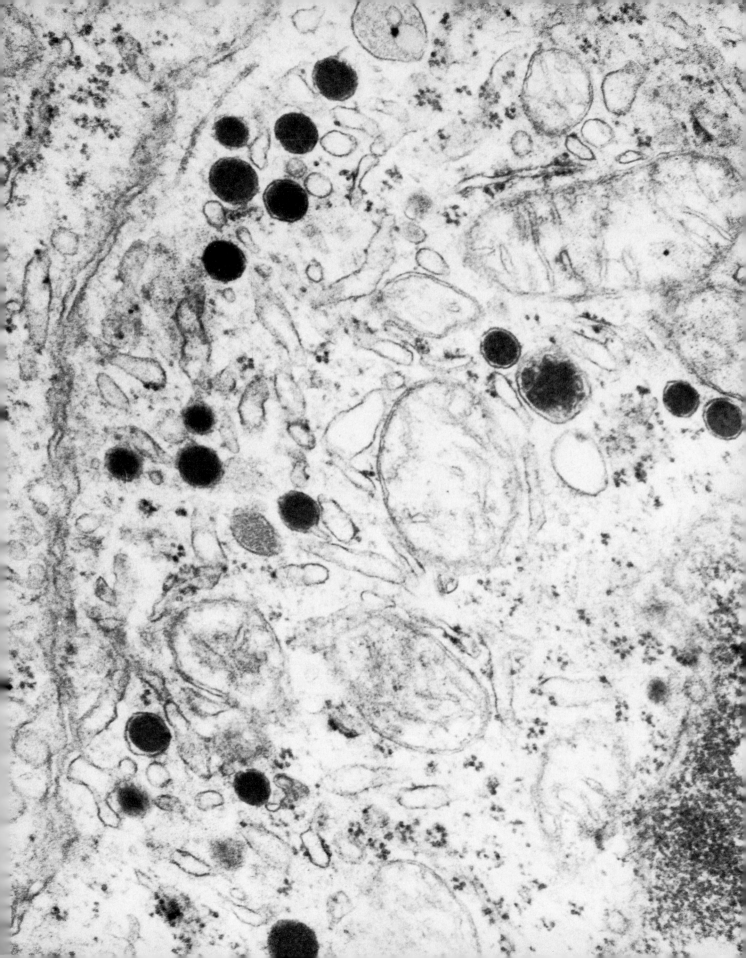

9 Thyroid Gland

Nodular Goiter

FIGURE 86. Parts of two thyroid follicles are surrounded by very well-formed basal lamina. The cytoplasmic membranes that are in contact with the colloid form several small microvilli. The rough endoplasmic reticulum appear to be dilated, and free ribosomes are scattered throughout the cytoplasm. Small Golgi complexes are also present. Lipid inclusions and small dense bodies are present. Cells have tight junctions and desmosomes. Follicles are separated by loose fibroconnective tissue containing blood vessels. (7150×)

172

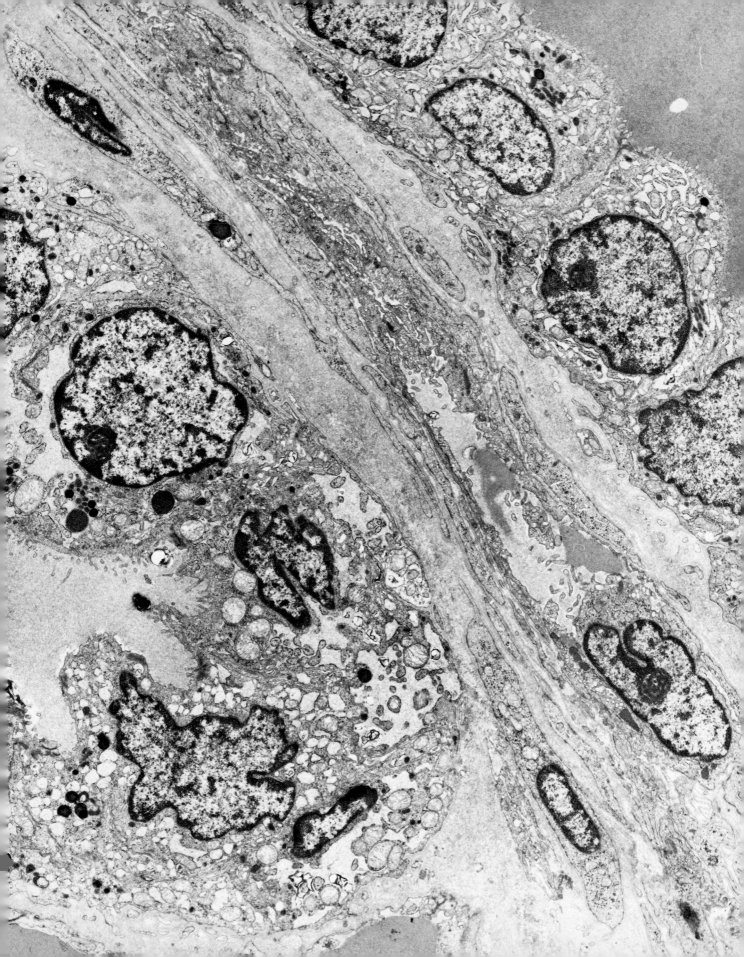

Follicular Carcinoma, Thyroid

FIGURE 87. This tumor is made of two components: a follicular and a solid one. The follicular portion consists of a single row of tall columnar cells with apical cytoplasm protruding above the two adjacent zonula occludens. The cytoplasm which is well polarized is packed with rough endoplasmic reticulum which appeared as a system of irregular shaped vesicles, sacs, and tubules whose lumens contain flocculent material. Mitochondria, dense bodies (probably lysosomes), glycogen particles are interspersed between the RER. The apicolateral cytoplasm display microvilli that protrude into the colloid which appears as a moderately electron dense material. The ultrastructure resembles that of an hyperplastic cell. (13,000×)

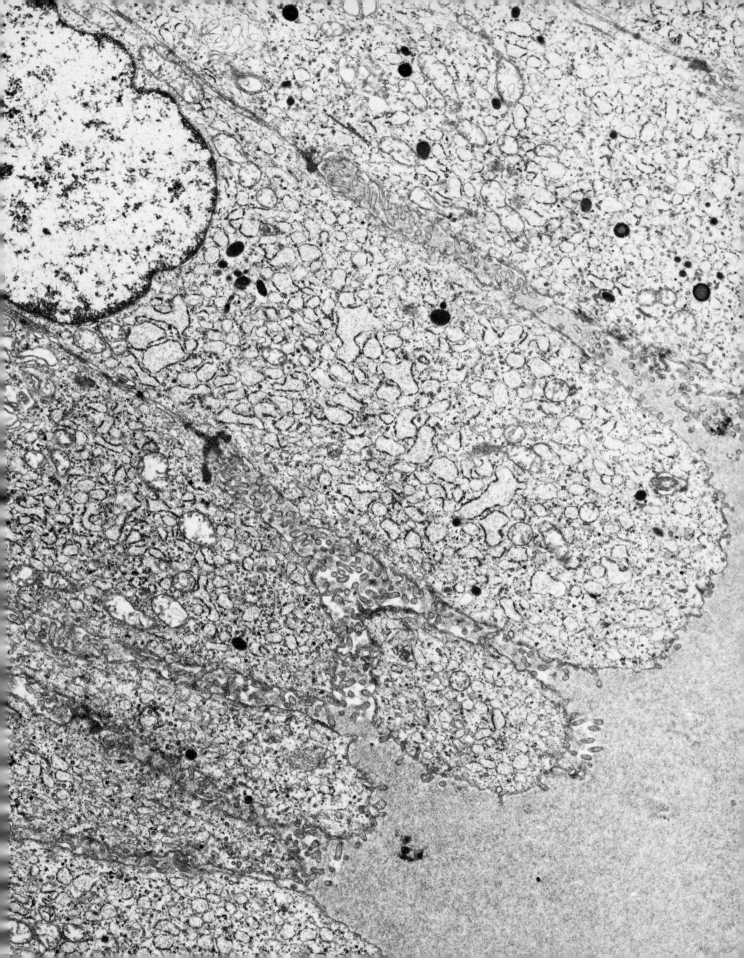

Follicular Carcinoma, Thyroid

FIGURE 88. The solid component of the tumor consists of pleomorphic cells with interdigitating processes. The cytoplasm is rich in RER which frequently forms lamellar arrays. The Golgi complex is well developed. Scattered unconspicuous organelles present are mitochondria, lysosomes, and vacuoles. The nucleus is generally convoluted and contains the euchromatin, nucleolus, and inclusion body. (10,000×)

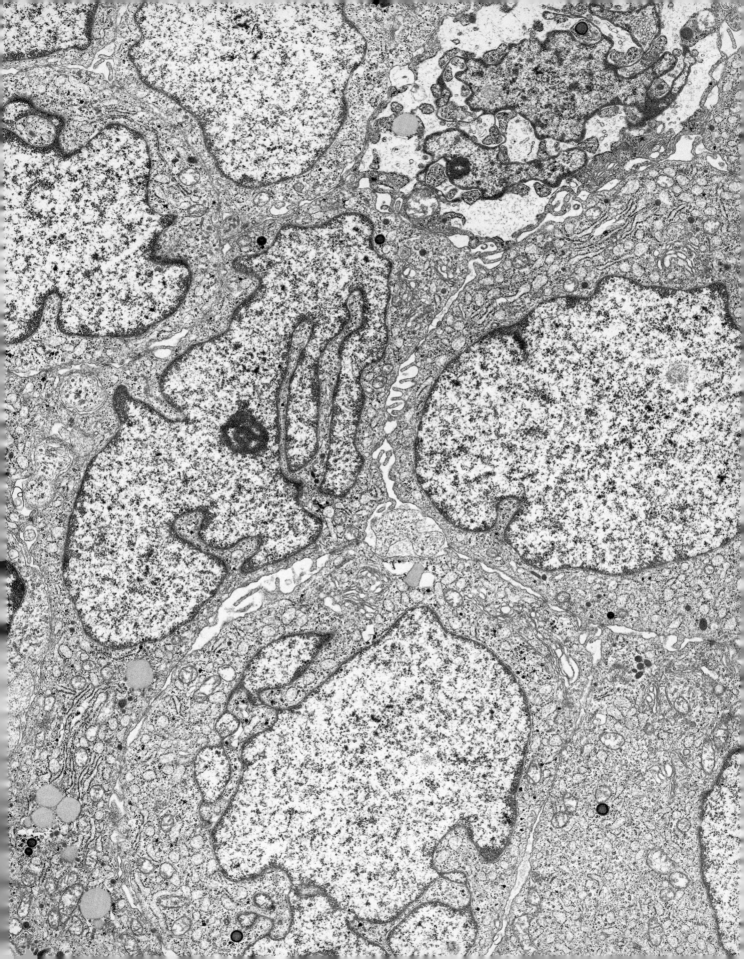

Follicular Carcinoma, Thyroid
(High Magnification, Basal Cytoplasm)

FIGURE 89.   This electron micrograph presents the basal portion of the neoplastic cells, lining the follicle. A delicate, continuous basal lamina is seen separating the tumor from the extra cellular space which contains a capillary partially surrounded by collagen fibrils. The dense vacuoles, apparently containing colloid material, are most numerous in the basal cytoplasm. (11,000×)

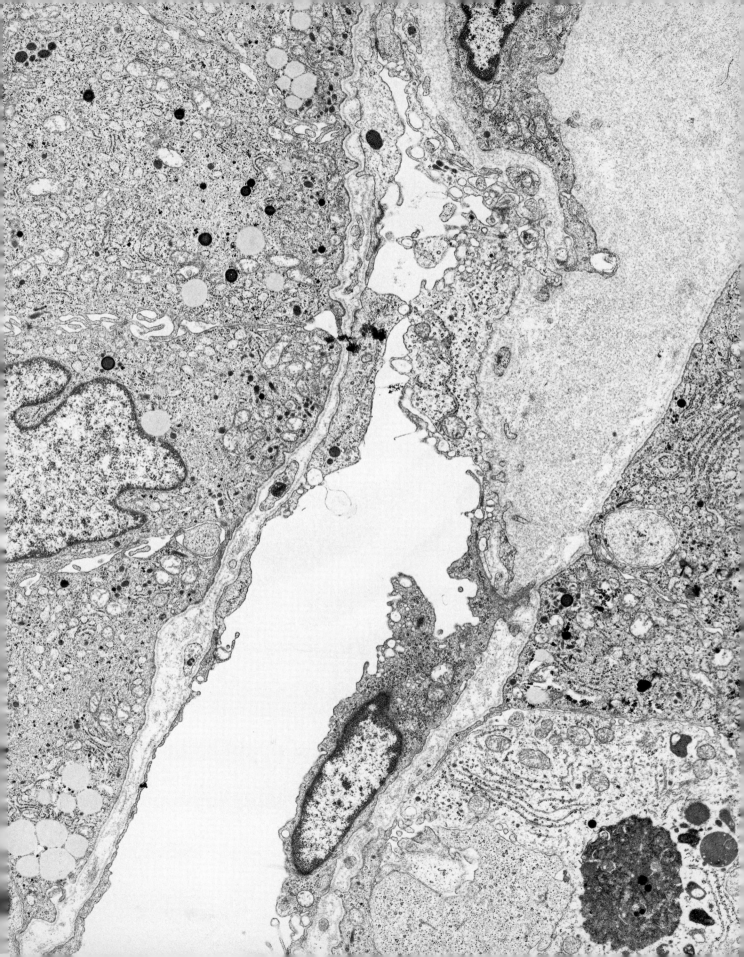

Follicular Carcinoma, Thyroid
(High Magnification, Solid Component)

FIGURE 90. This higher magnification of the solid component shows the lamellar array of RER in centrally located tumor cells. One neoplastic cell with convoluted nucleus contains several dense intracytoplasmic vacuoles, probably representing a colloid. A poorly differentiated cell is rich in free ribosomes. This cell is probably synthesizing protein for its own metabolism. (24,100×)

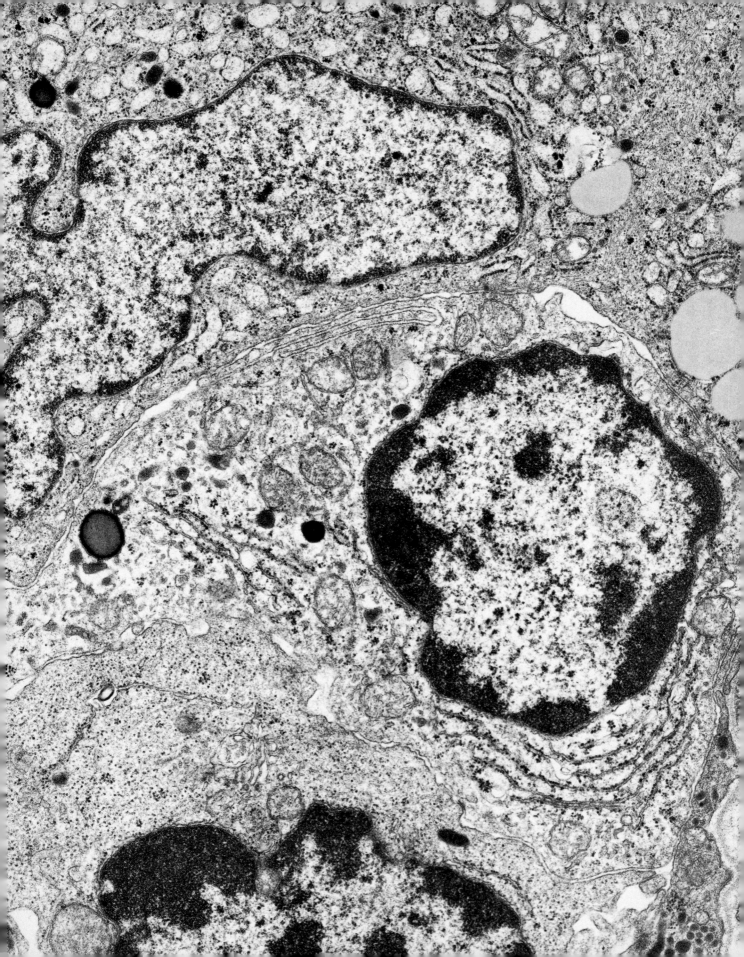

10 Parathyroid Gland

Parathyroid Adenoma (**Chief Cell Type**)

FIGURE 91. This electron micrograph demonstrates clusters of chief cells some of which show lipid droplets. The cells are arranged around a capillary. (9900×)

182

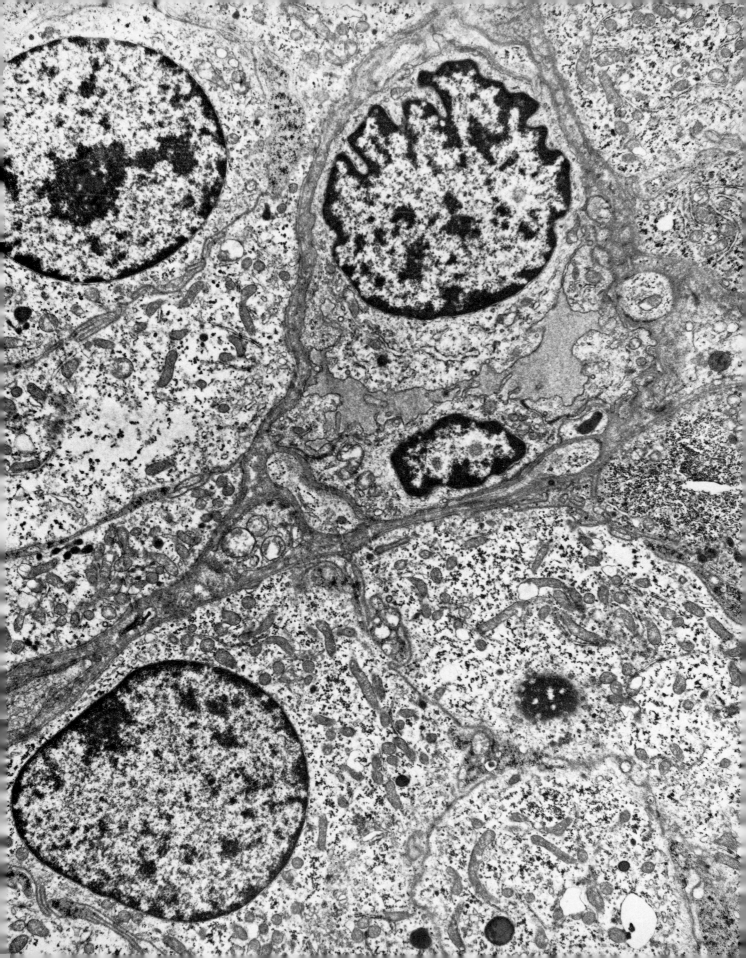

Intracytoplasmic Lumina in Tumors (Parathyroid Adenoma)

FIGURE 92.   An intracytoplasmic lumen with microvilli is clearly demonstrated in this picture. This intracellular lumen communicates with a main lumen or intracellular extension of intercellular spaces. Intracytoplasmic lumina formation has been reported in several epithelial tumors, such as ones from the breast, prostate, thyroid, and kidney. (15,800×)

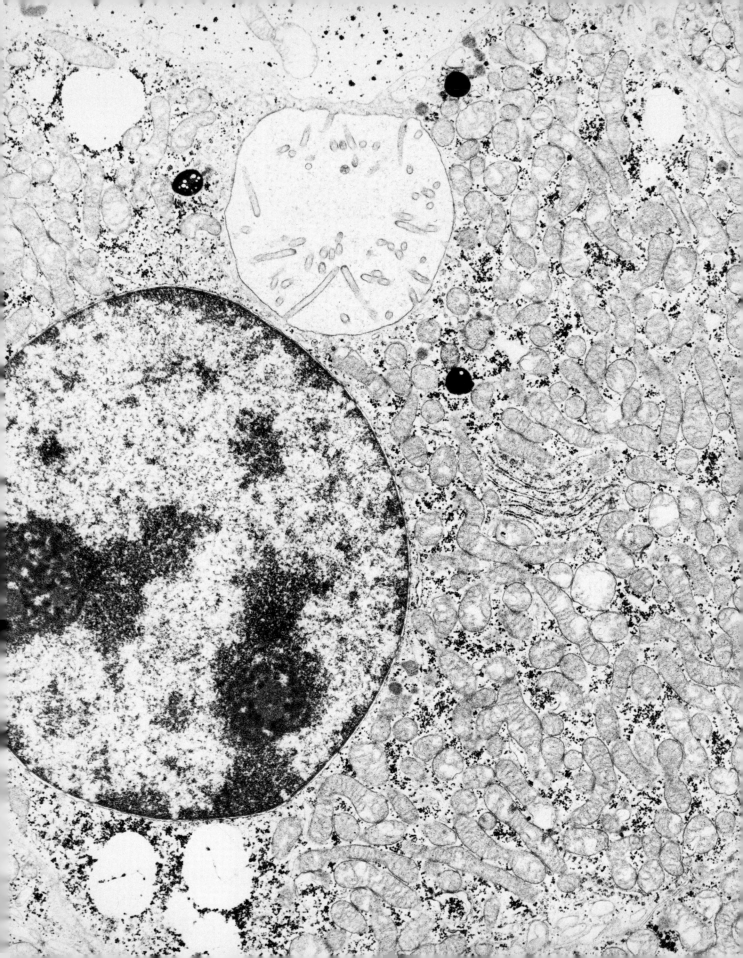

Parathyroid Adenoma (Oxyphil Cell Type)

FIGURE 93. Oxyphil cells are characterized by numerous closely packed mitochondria. Glycogen particles, ribosomes, and several small round electron dense granules are interspersed between the mitochondria. The rough endoplasmic reticulum is aggregated into multiple lamellar or whorl-like arrays. Occasionally the neoplastic or hyperplastic parathyroids are composed almost exclusively of oxyphil cells. Oxyphil cells in functional adenomas and parathyroid hyperplasia frequently contain secretory granules, aggregates, and dispersed sacs of endoplasmic reticulum, as well as Golgi apparatus (not shown here). (7200×)

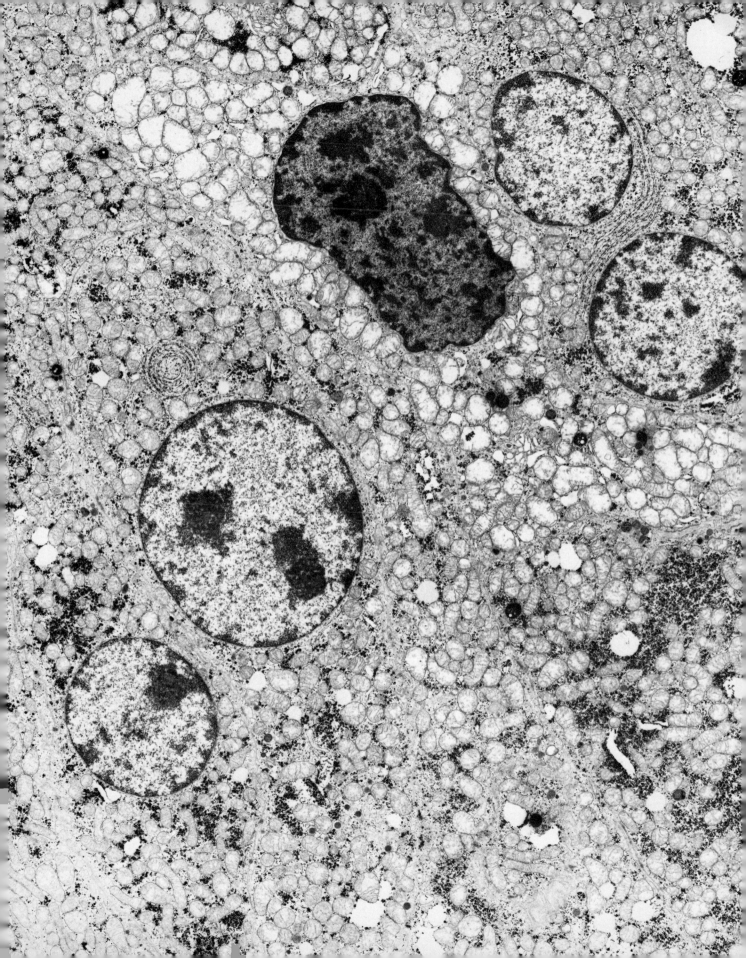

Parathyroid Adenoma (Oxyphil Cell Type)

FIGURE 94. A capillary endothelial cell shows multiple fenestrae in the cytoplasm and has electron dense bodies, resembling secretory granules within the cytoplasmic projection. The surrounding oxyphil cells also have many secretory granules (arrow) in the basal cytoplasm of the cells, i.e. toward the capillary.

This finding may represent secretory material being transported to the capillary lumen, although this remains to be proven conclusively. (9600×)

188

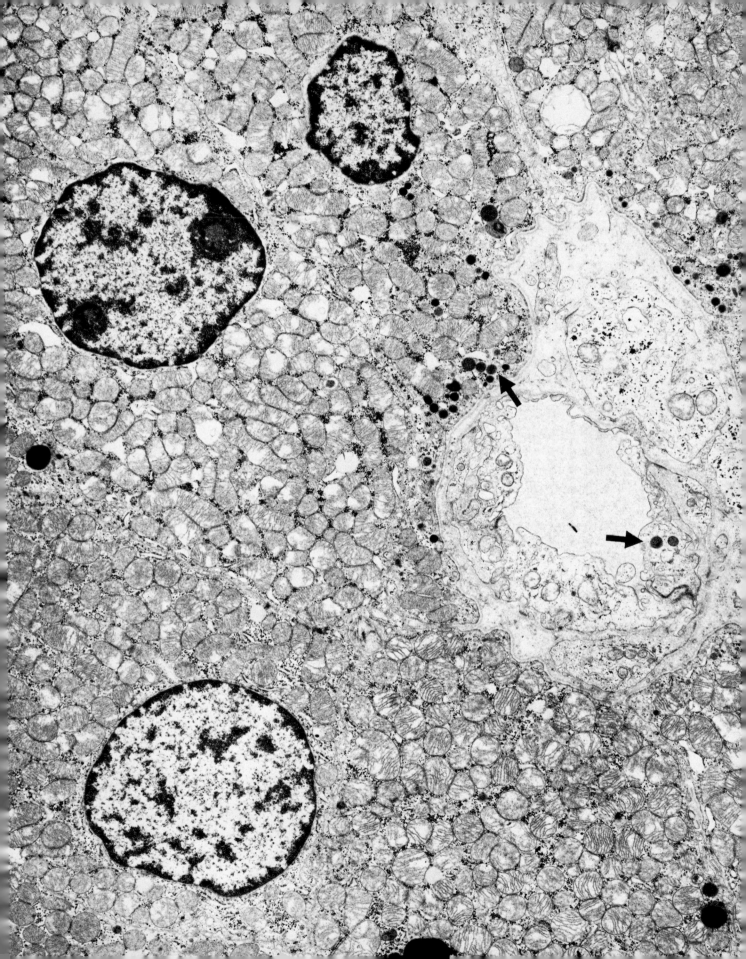

Parathyroid Adenoma (Oxyphil Cell Type)

FIGURE 95. High magnification of an oxyphil cell shows numerous mitochondria. (31,900×)

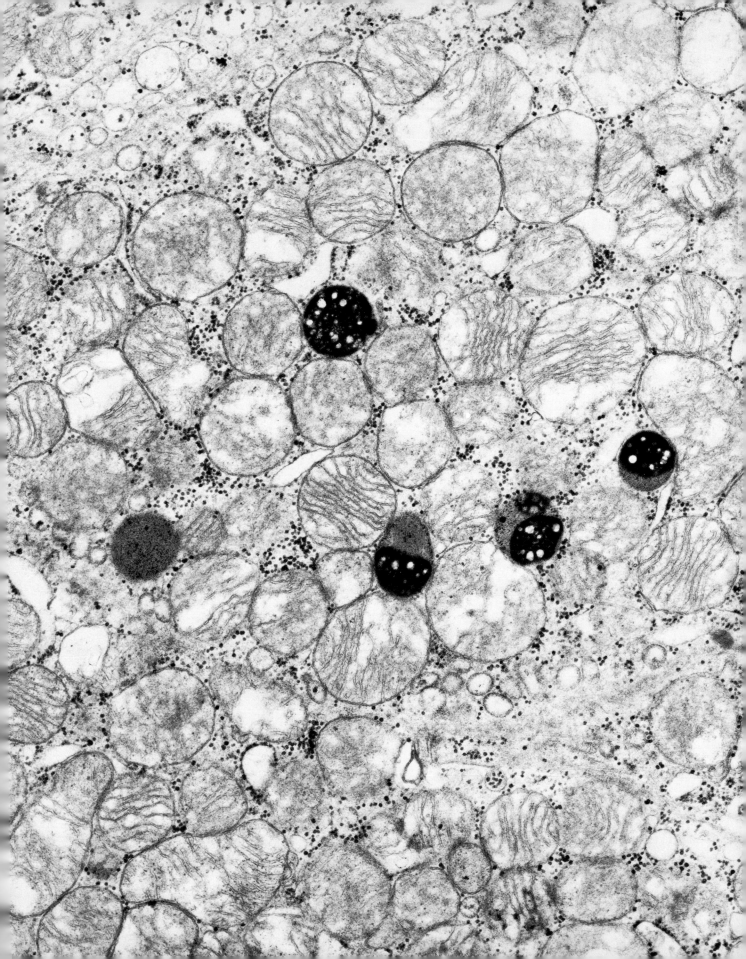

Parathyroid Adenoma

FIGURE 96. Portion of a chief cell and an oxyphil cell are seen. The oxyphil cell cytoplasm is rich in mitochondria. The mitochondrial cristae are abundant, and large mitochondrial granules are present. (27,100×)

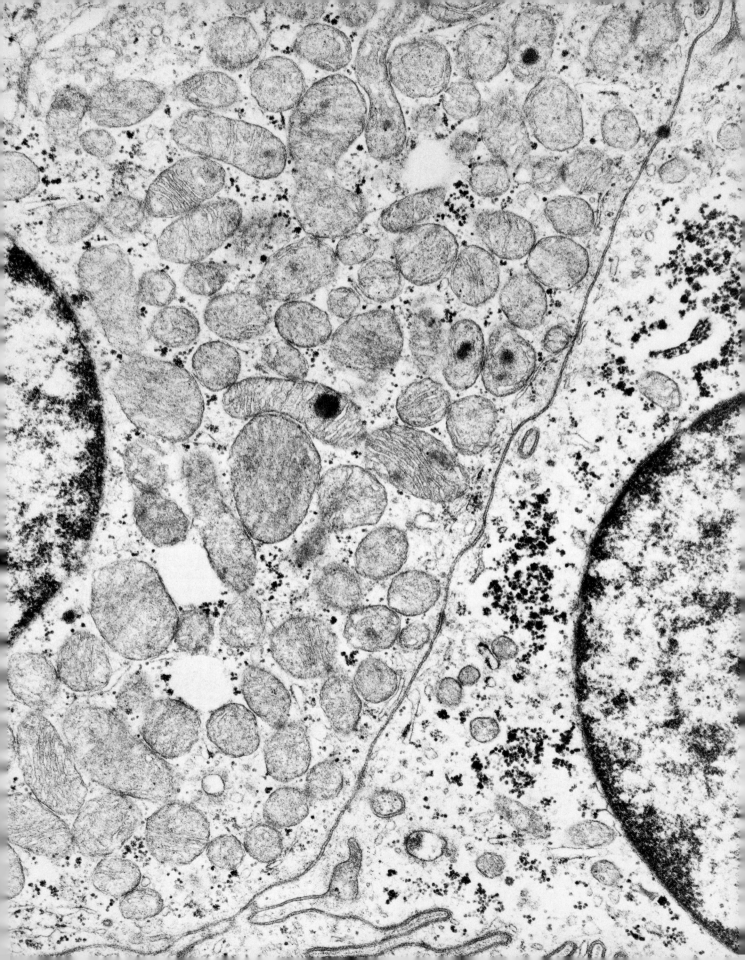

11 Thymus

Thymoma, Mixed Type

FIGURE 97. The electron micrograph depicts the epithelial and lymphocytic components of mixed-type thymoma.

The lymphocyte, indicated by the thick arrow, has an oval nucleus with heavy chromatin condensation along the nuclear membrane. The cytoplasm is scanty and contains a few mitochondria and many free ribosomes. The epithelial cell is characterized by desmosomes which are indicated by thin arrows. (20,400×)

104

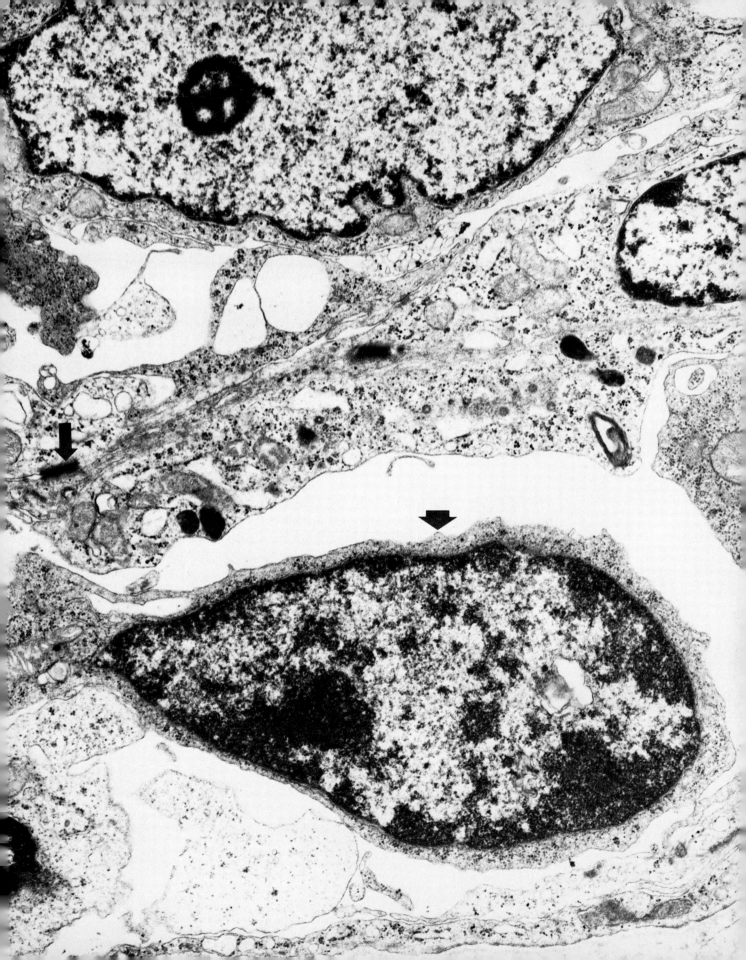

12 Chemoreceptor System

Carotid Body Tumor

FIGURE 98. The tumor cells are closely packed together and the cytoplasm is rich in organelles. They are intimately related to the capillary walls from which they are separated by a thin rim of collagen. ($8580\times$)

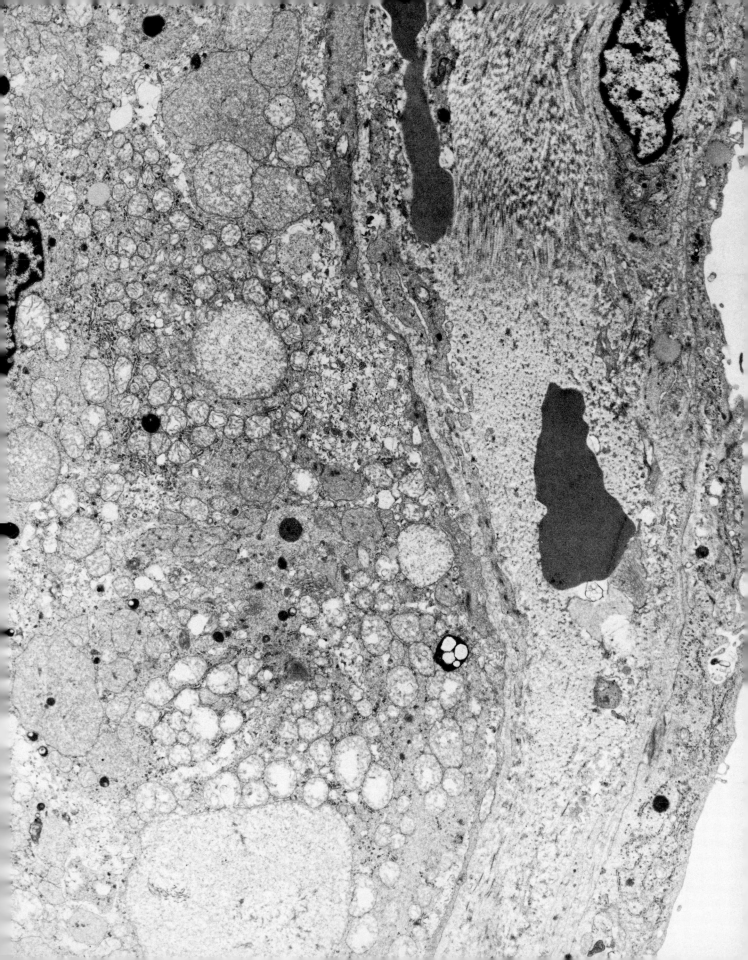

Carotid Body Tumor (Higher Magnification)

FIGURE 99. The tumor cells, usually polygonal in shape, demonstrate closely apposed borders with randomly scattered true desmosomal junctions. Mitochondria are numerous, and many are swollen. There are stacks of smooth endoplasmic reticulum in a perinuclear location and many randomly dispersed neurosecretory types of granules. (10,100×)

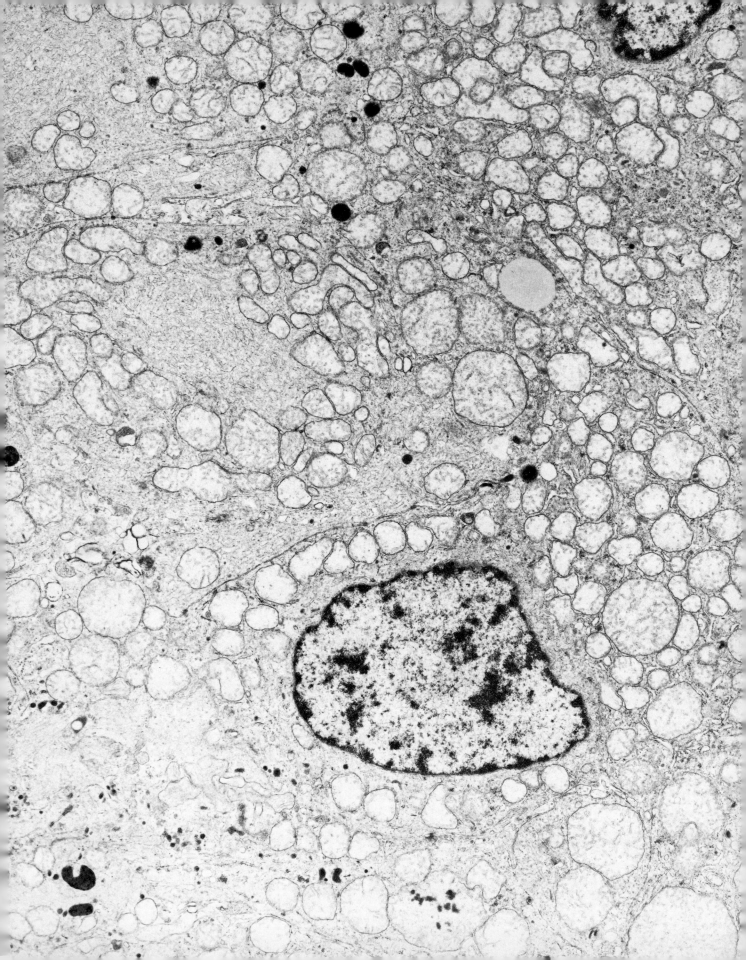

Neurosecretory Granules of a Carotid Body Tumor

FIGURE 100. The cells exhibit two distinct populations of granules distinguished on the basis of size and configuration. The most prominent ones are the small round granules with extremely dense cores surrounded by a distinct membrane. The second type are large, round, but occasionally oblong and have granules with a moderately electron dense core which is usually homogeneous but occasionally contains inclusions. (39,800$\times$)

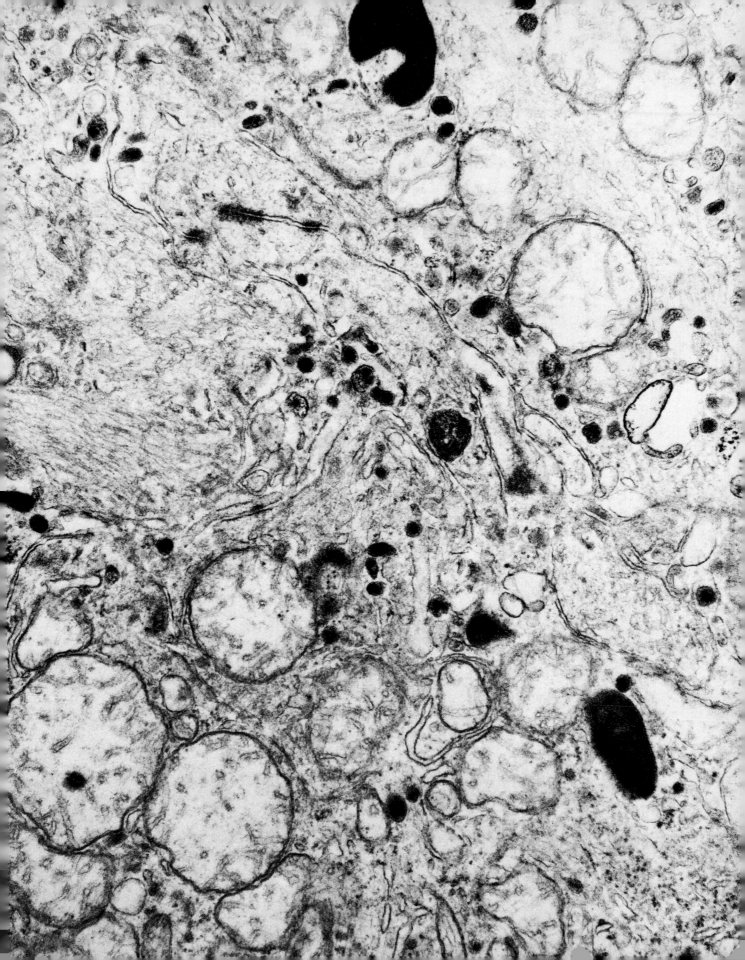

13 Adrenal Gland

Adenomatoid Tumor of the Adrenal Gland

FIGURE 101. The electron micrograph shows part of a luminal space lined by a single layer of flattened cells. The cells have numerous microvillous projections, are connected by desmosomes, and rest on a well-formed basal lamina (not shown in the picture). Another feature of the tumor cell is the haphazard cytoplasmic distribution of bundles of wavy, filamentous structures resembling tonofilaments. The luminal spaces, numerous microvilli and filamentous structures resembling tonofilaments strongly suggest and support an origin of adenomaotoid tumor from mesothelial cells. We wish to note that the location of this adenomatoid tumor in the adrenal glands is extremely rare and to our knowledge previously unreported. (27,650×)

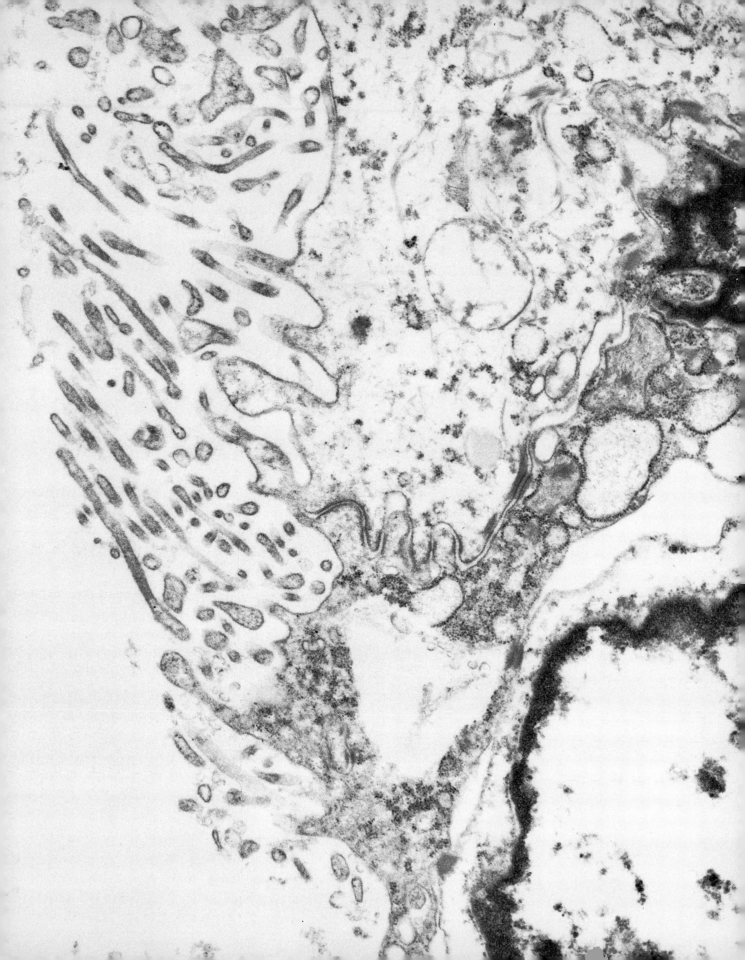

14 Kidney

Berger Nephropathy

FIGURE 102.   This electron micrograph shows a part of a glomerulus from a 12-year-old female with recurrent microscopic hematuria and mild proteinuria. Note the marked proliferation of mesangium with prominent immune complex deposits. The glomerular capillary basement membranes remain fairly normal. There is focal fusion of epithelial foot processes. On immunofluorescent study there is generalized reaction for I_gA, I_gG, and β_1C in the mesangium. Berger nephropathy is not an uncommon nephritis in childhood; hematuria, either microscopic or gross, with relatively normal renal function is common clinical presentation. The prognosis is generally favorable. The presence of I_gA in the mesangium is characteristic for this lesion. ($9600\times$)

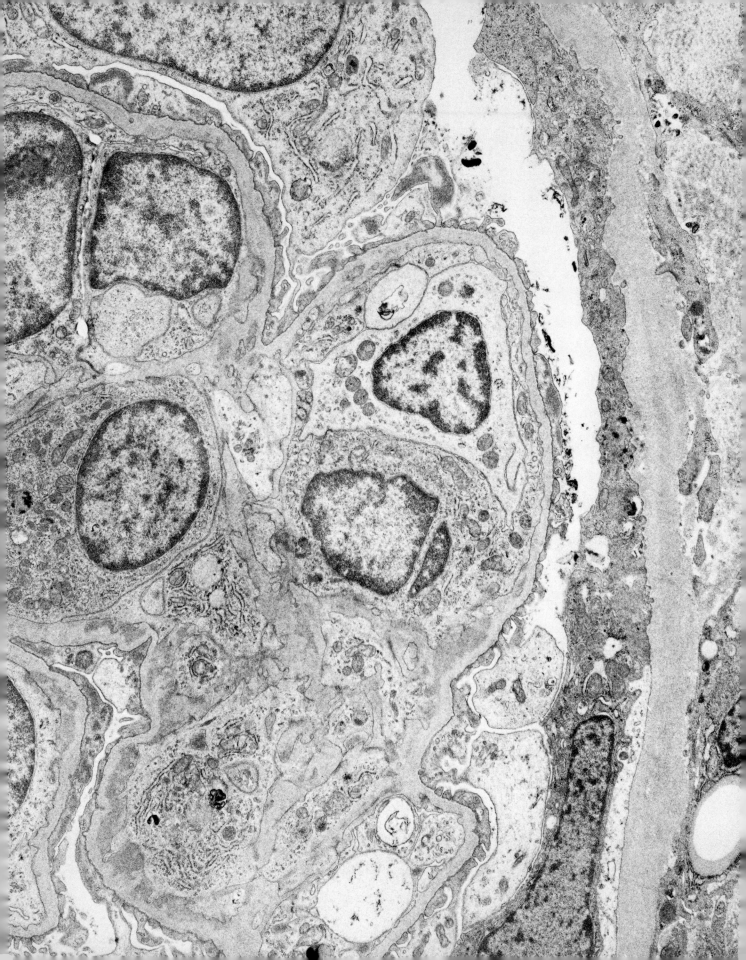

Acute Postinfectious Glomerulonephritis

FIGURE 103.   This electron micrograph is taken from a renal biopsy of a 14-year-old girl with delayed recovery from acute post-streptococcal glomerulonephritis. Note the proliferation of mesangial cells and the sub-epithelial hump which are characteristic for this lesion. Glomerular basement membranes are somewhat wrinkled but do not appear thickened, and capillary lumens are patent. Acute post infectious glomerulonephritis is usually associated with group A streptococcal infection, but occasionally pneumococcal, staphylococcal, and other bacterial infections are found to be the cause. Although delayed recovery is not rare, complete recovery with few exceptions is the rule. The disease usually affects children and young adults. (11,400×)

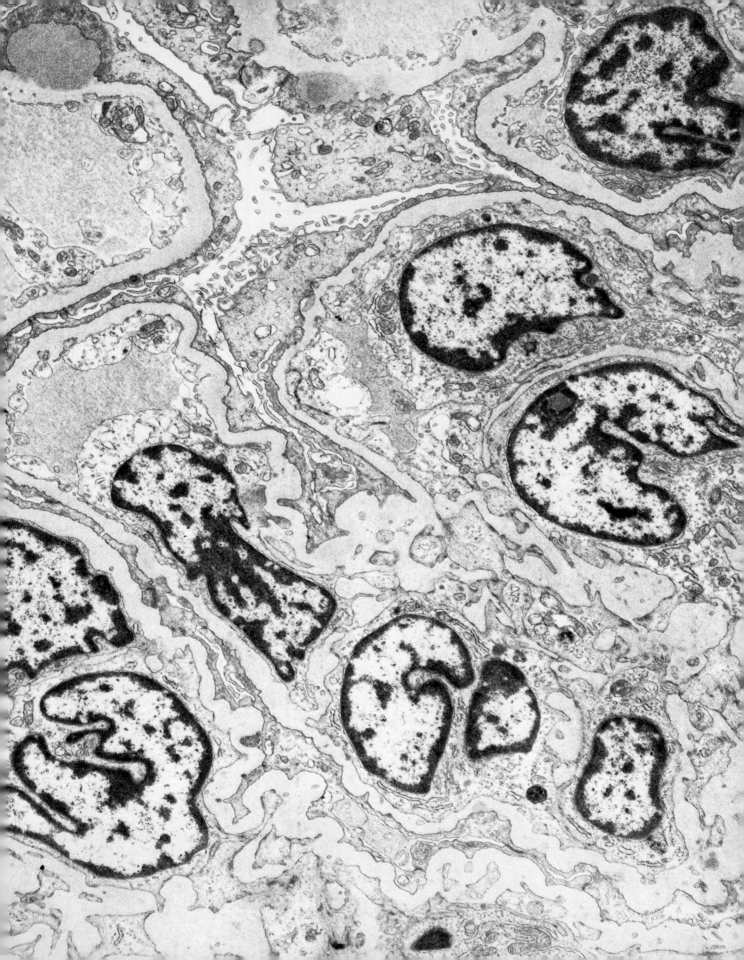

**Subepithelial Hump in Acute
Postinfectious Glomerulonephritis**

FIGURE 104. This high magnification of the preceding photograph shows electron dense material accumulated in the subepithelial side of the glomerular basement membrane. (32,400×)

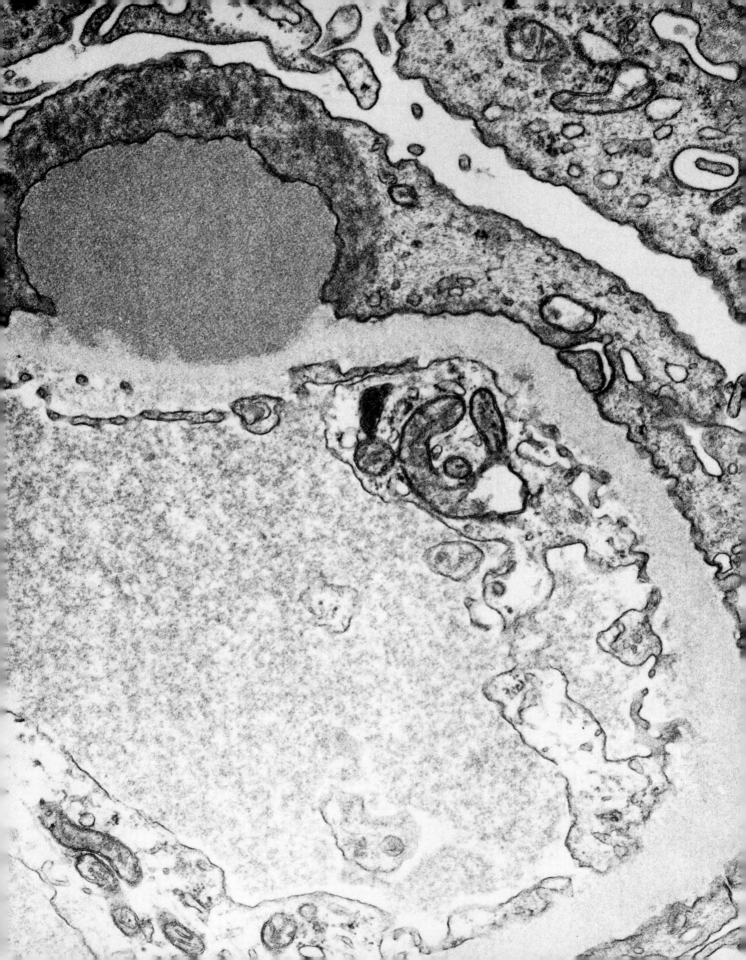

**Postinfectious Glomerulonephritis Superimposed
On Diabetic Glomerulosclerosis**

FIGURE 105. The usual immune complex "hump" of postinfectious glomerulonephritis is seen on the epithelial aspect of the basement membrane. The mesangium shows increased matrix. (13,800×)

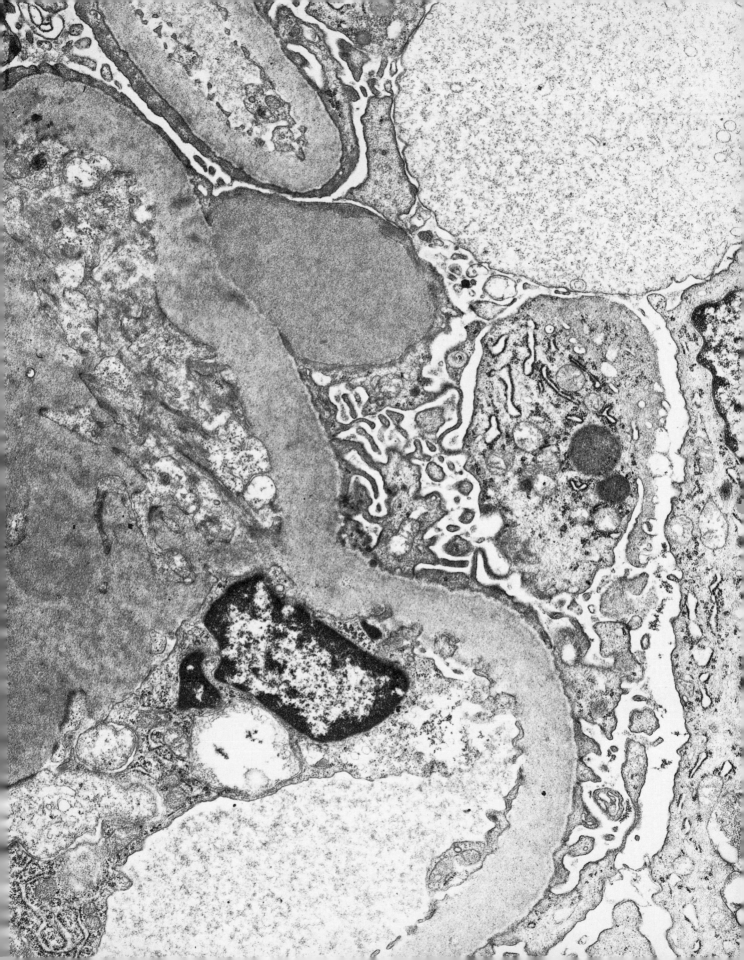

**Diabetic Glomerulosclerosis,
Kimmelstiel-Wilson's Disease**

FIGURE 106. The glomerular loops show mild focal thickening of the basement membrane. The foot processes are discrete. The most striking finding is the presence of increased basement membrane-like material arranged in an irregular anastomosing network in the mesangium giving the lobule a nodular appearance. (7400×)

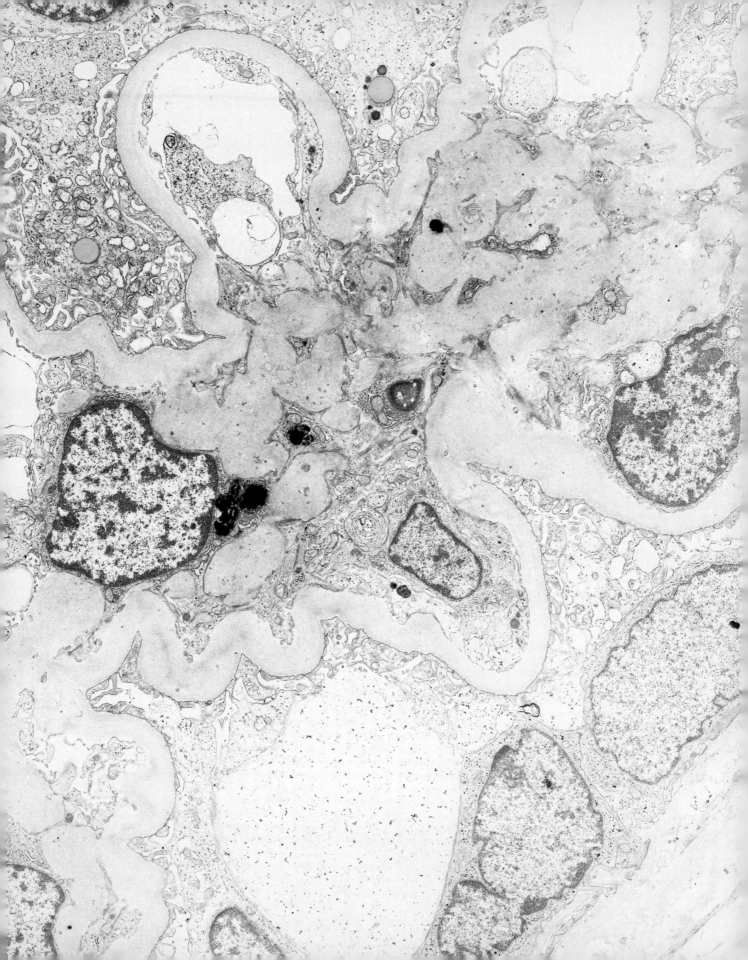

Lupus Nephritis with Minimal Glomerular Alteration

FIGURE 107.   This is a portion of a glomerulus from a 14-year-old male who has mild proteinuria and systemic lupus erythematosus. Note that the basement membrane remains fairly uniform and that the epithelial foot processes are discrete. Immune complexes or dense deposits (D) are occasionally seen in the mesangium. Undulating tubular aggregates (virus-like structures) are seen in dilated endoplasmic reticulums of the endothelial cells (arrows). (15,300×)

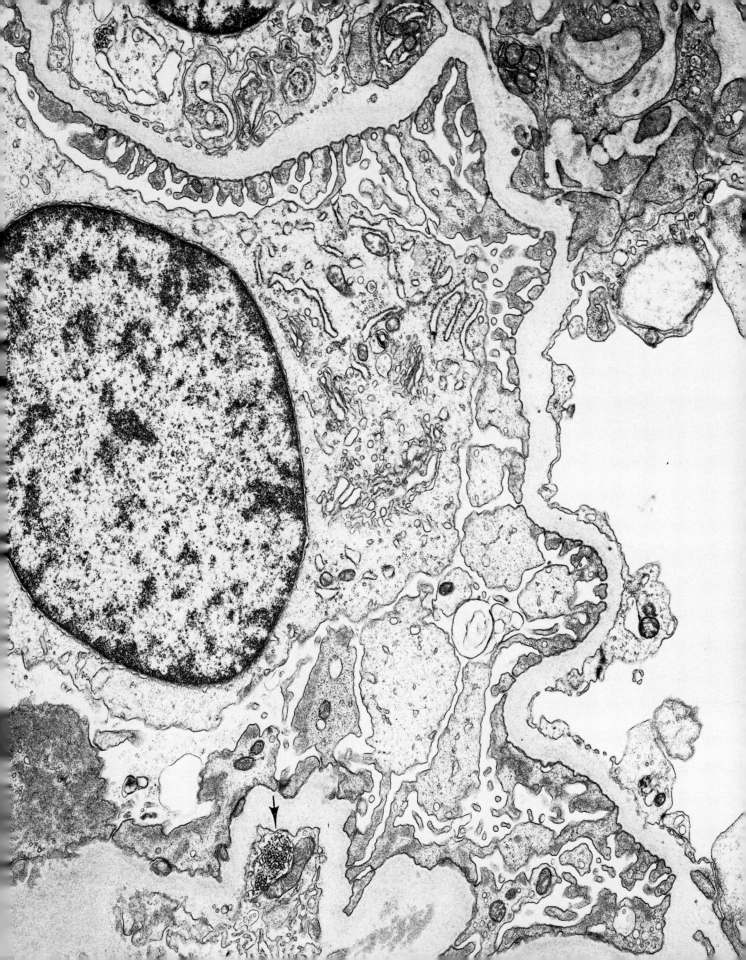

Kidney

Lupus Nephritis with Minimal Glomerular Alteration

FIGURE 108. This higher magnification of the preceding picture shows the undulating tubular aggregates (arrow) in the dilated endoplasmic reticulum of an endothelial cell. (31,200×)

216

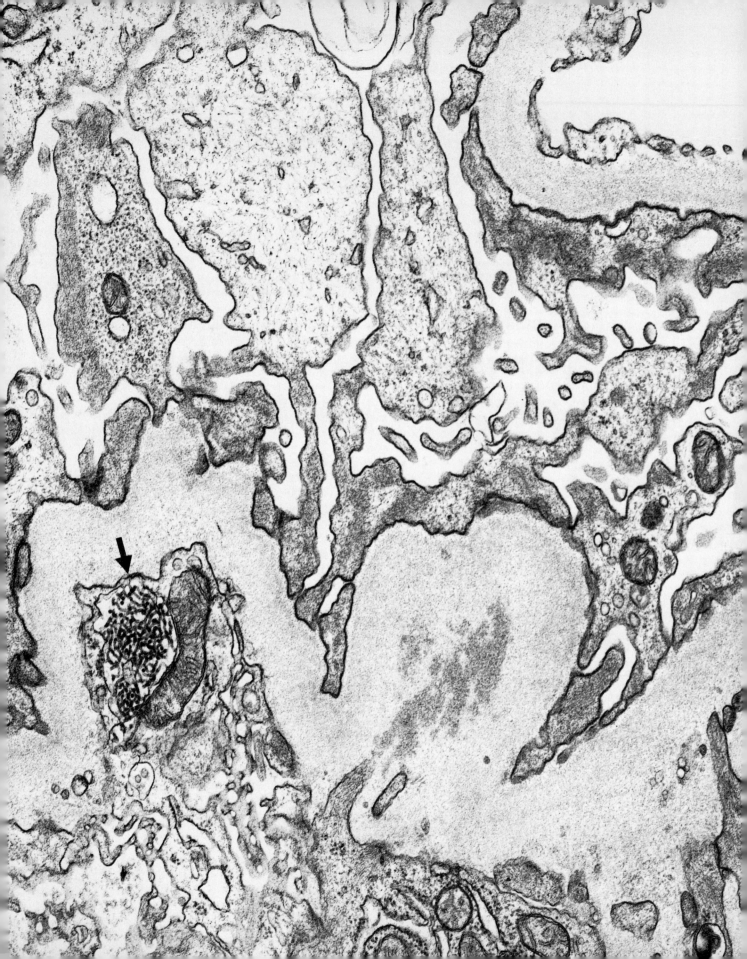

Lupus Nephritis

FIGURE 109. There are massive, diffuse osmiophilic deposits (immune complexes) occupying the mesangial areas and subendothelial aspects of the glomerular basement membranes. (4500×)

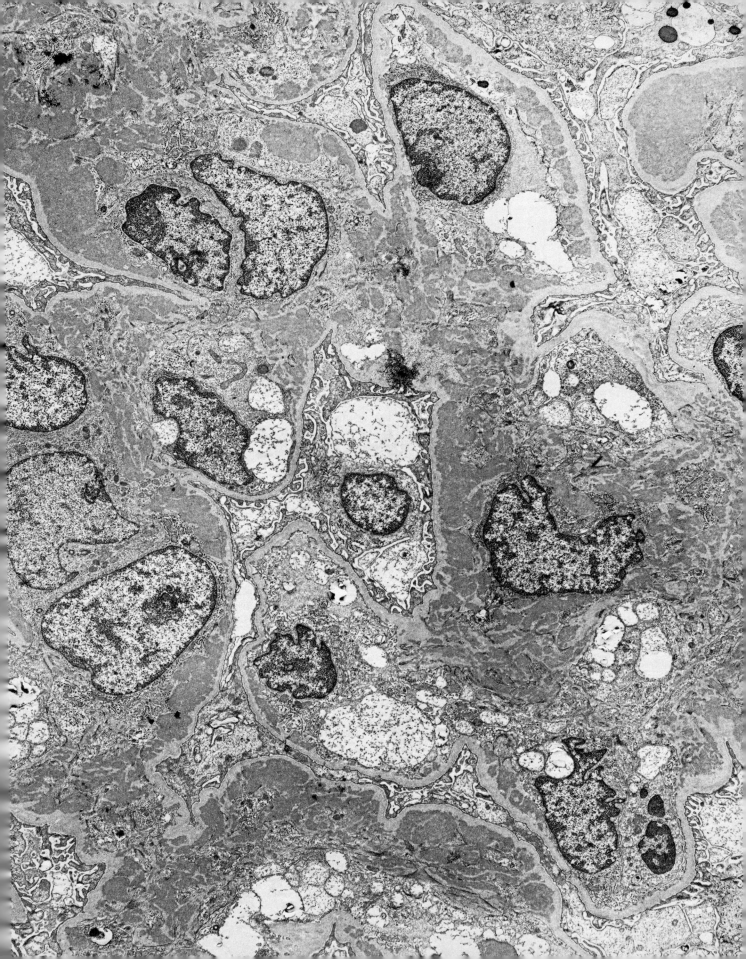

Lupus Nephritis

FIGURE 110. The subendothelial aspect of glomerular basement membrane contains electron-dense deposits. Viral-like particles are within the endothelial cell cytoplasm. (26,400×)

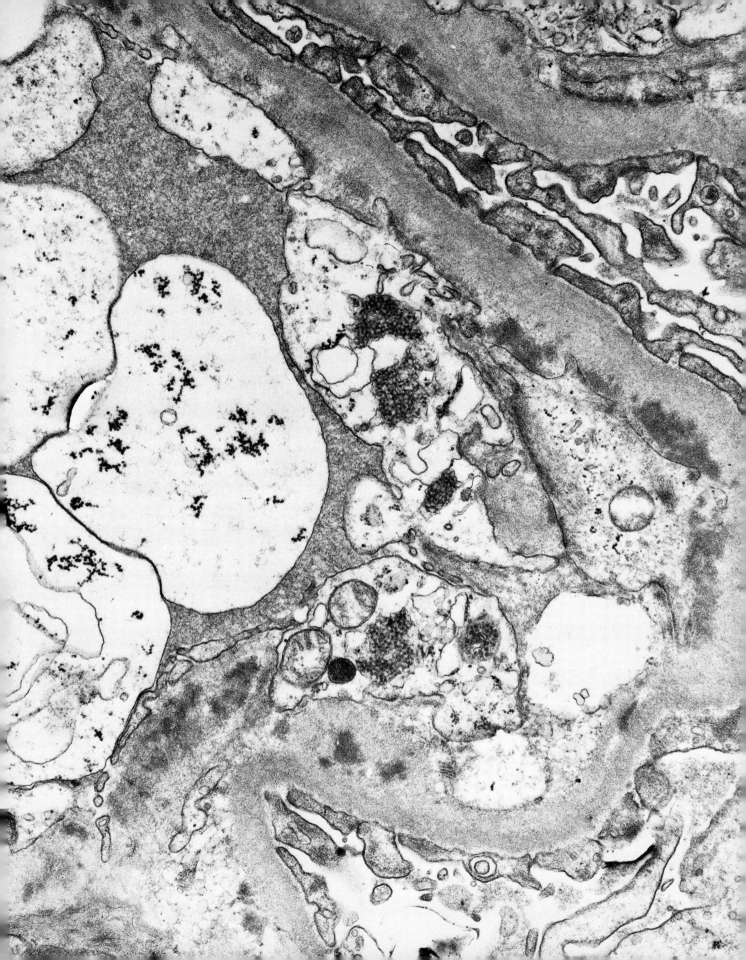

Lupus Nephritis

FIGURE 111. This is an electron micrograph of membranous nephropathy in a patient with systemic lupus erythematosus. The osmiophilic dense deposits (immune complexes) are located along the epithelial aspect of the glomerular basement membrane. The epithelial cell foot processes are approximated (fused) in the involved segments. (27,300×)

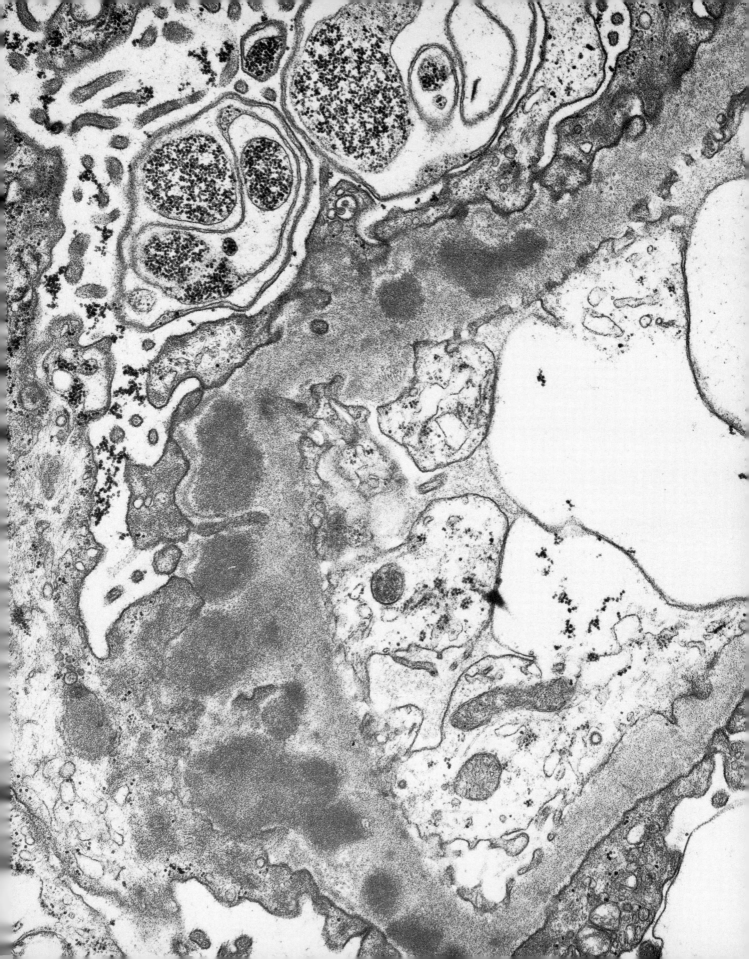

Lupus Nephritis

FIGURE 112. This electron micrograph shows an organized paracrystalline deposit in the glomerular basement membrane. (17,000×)

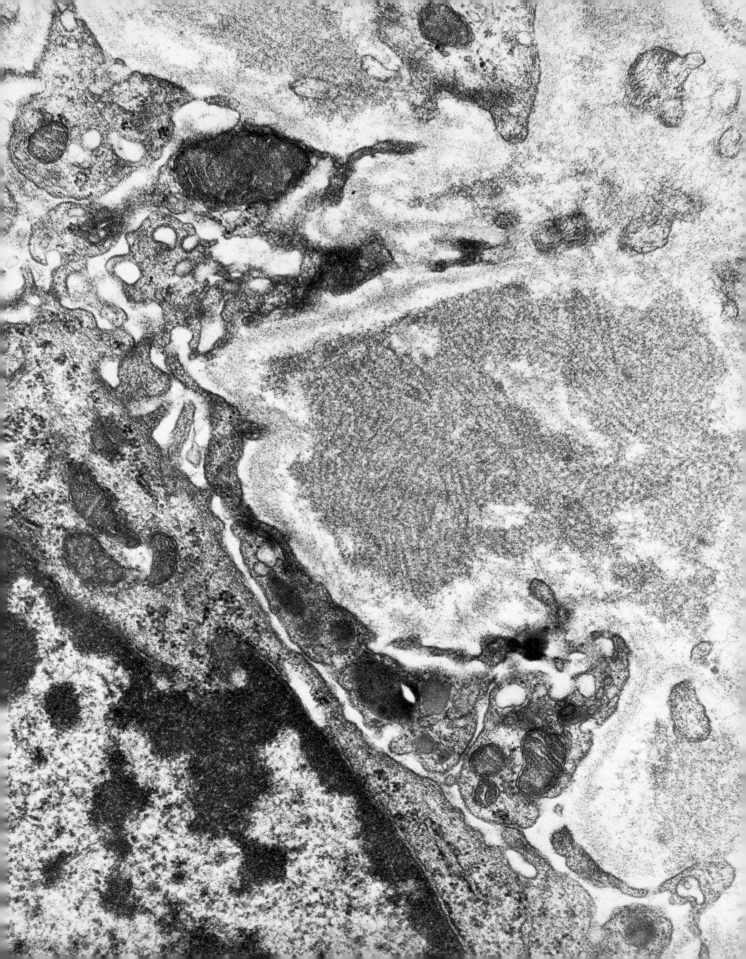

Early Membranous Nephropathy

FIGURE 113. This micrograph demonstrates the few and small immune complex deposits associated with segmental approximation of the epithelial cell foot processes. (27,100×)

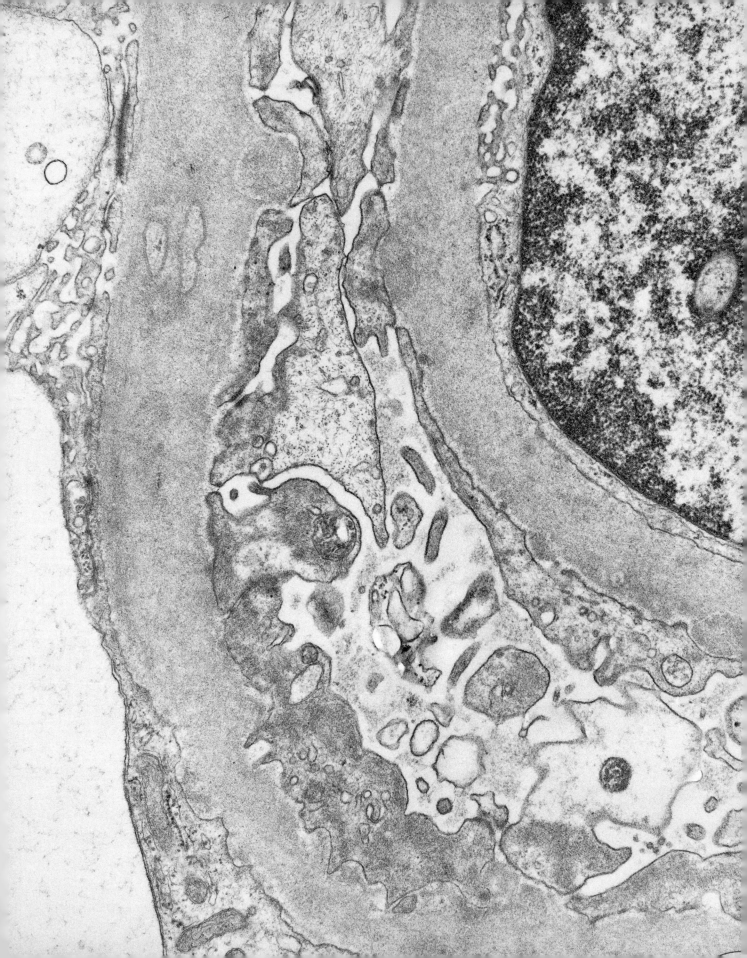

Membranous Nephropathy

FIGURE 114. The epithelial aspect of the glomerular basement membrane contains electron-dense deposits. (35,200×)

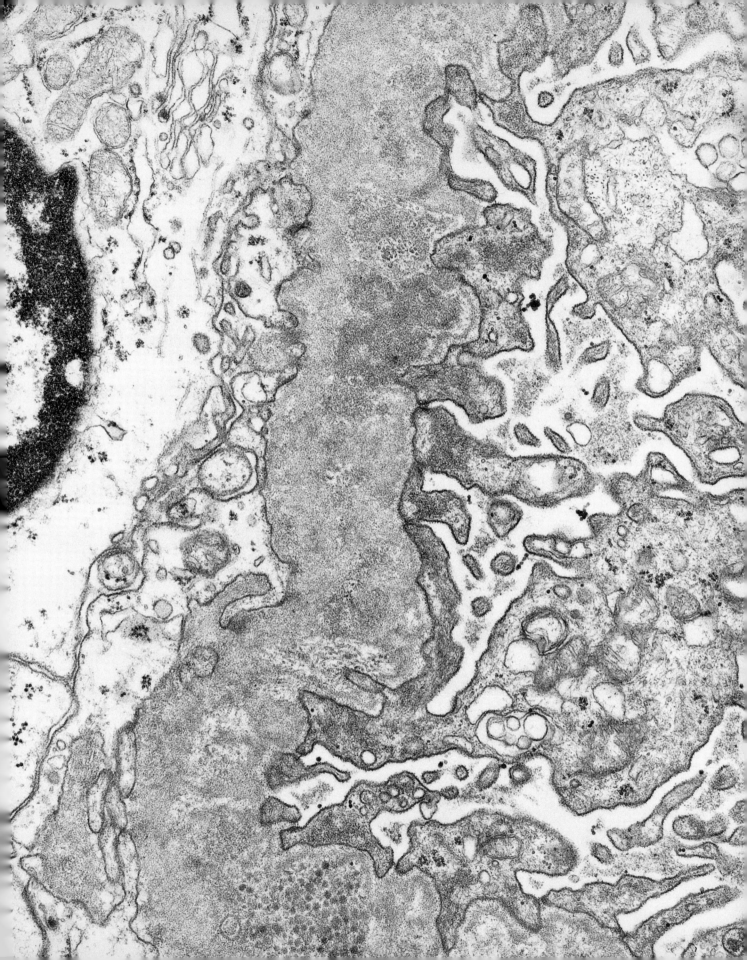

Membranous Nephropathy (Advanced)

FIGURE 115. The glomerular basement membrane is markedly thickened, and original sites of deposits are relatively electron-lucent. Epithelial cell foot processes are segmentally approximated. (3800×)

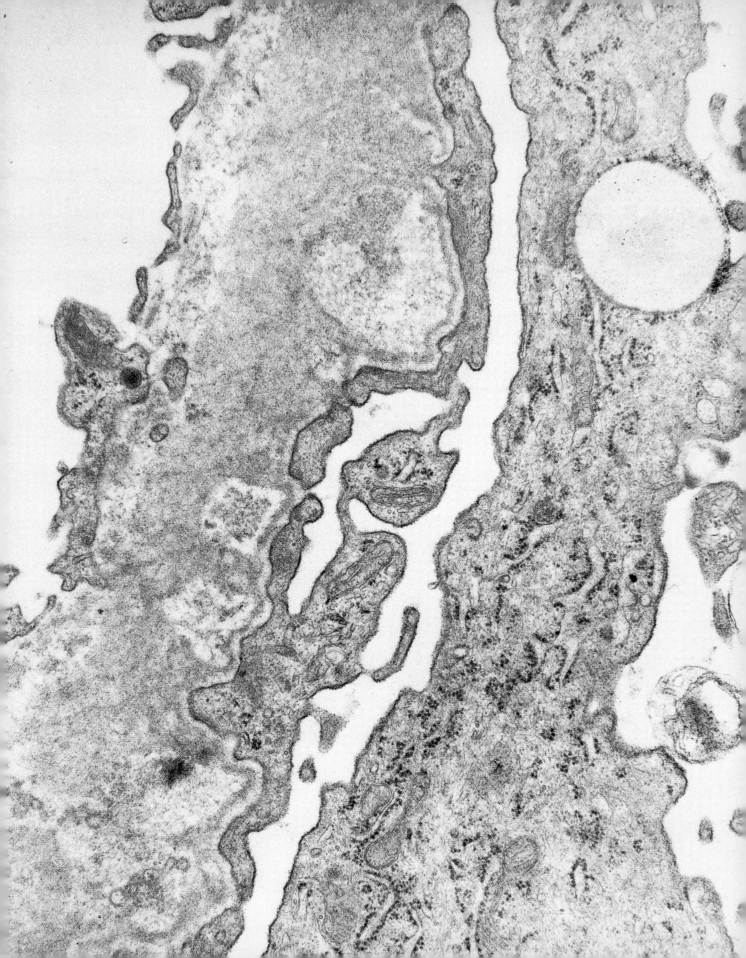

Hereditary Nephritis (Alport's Syndrome)

FIGURE 116. This is a portion of a glomerular tuft from an 11-year-old male who presented with hematuria and mild proteinuria. Note the irregularity and thickening of the basement membrane and the splitting of lamina densa (arrows). Although the light microscopic picture varies from case to case, the fine structural change of the basement membrane is characteristic for this lesion. Microscopic hematuria is a constant finding, and a high frequency hearing defect is noticed in 30% of the patients. It is a disorder with an autosomal dominant inheritance. (30,900×)

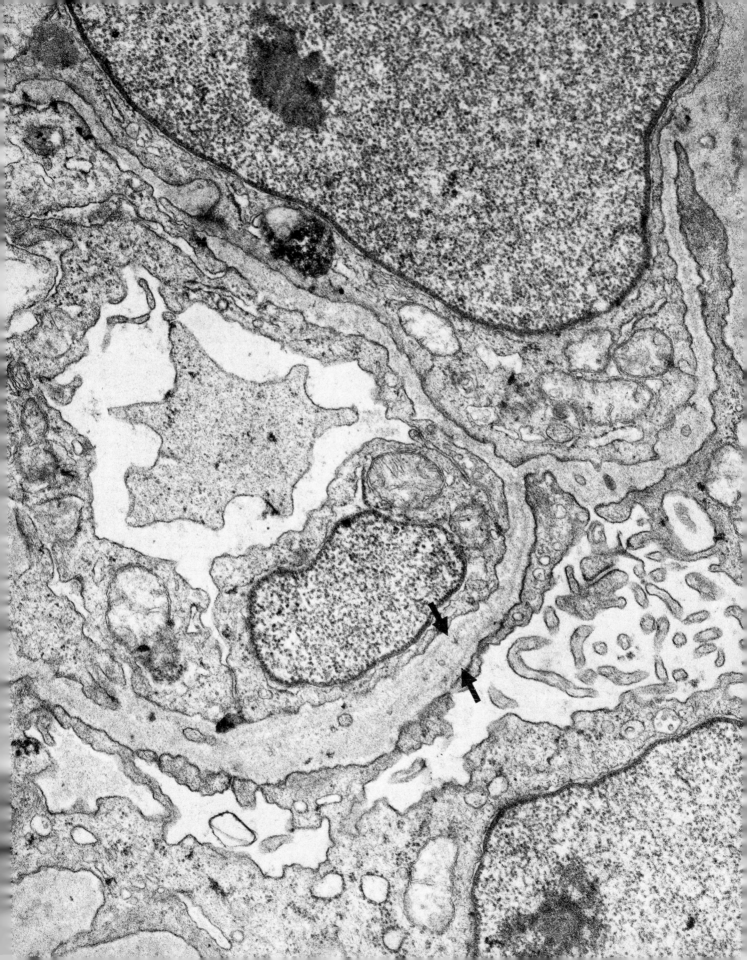

Hereditary Nephritis (Alport's Syndrome)

FIGURE 117. This is a high magnification of the glomerular basement membrane from the preceding case. Note the splitting of lamina densa. (53,700×)

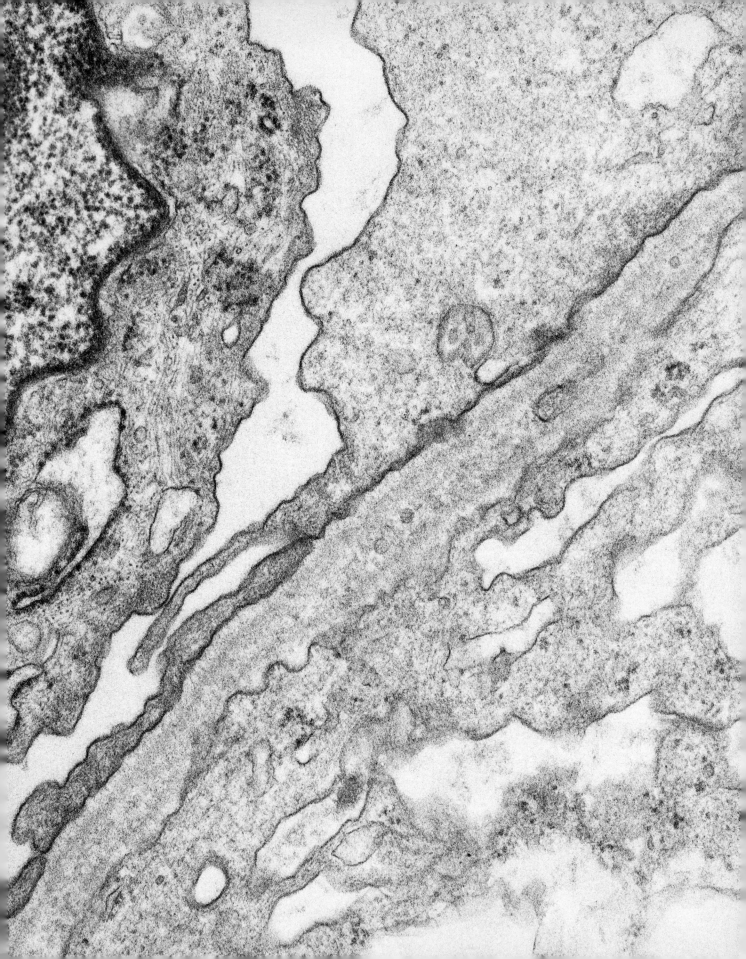

Microangiopathic Nephropathy

FIGURE 118. This is a portion of a glomerular tuft from a female young-ster who recovered from a recent episode of "hemolytic uremic syndrome." Note that the glomerular capillary lumen is partially obliterated and that there is occasional residual fibrin aggregate (arrow) in the subendothelial side of the basement membrane. The basement membrane is thickened and there is abundant collagen fibers within the basement membrane-like material. (53,700×)

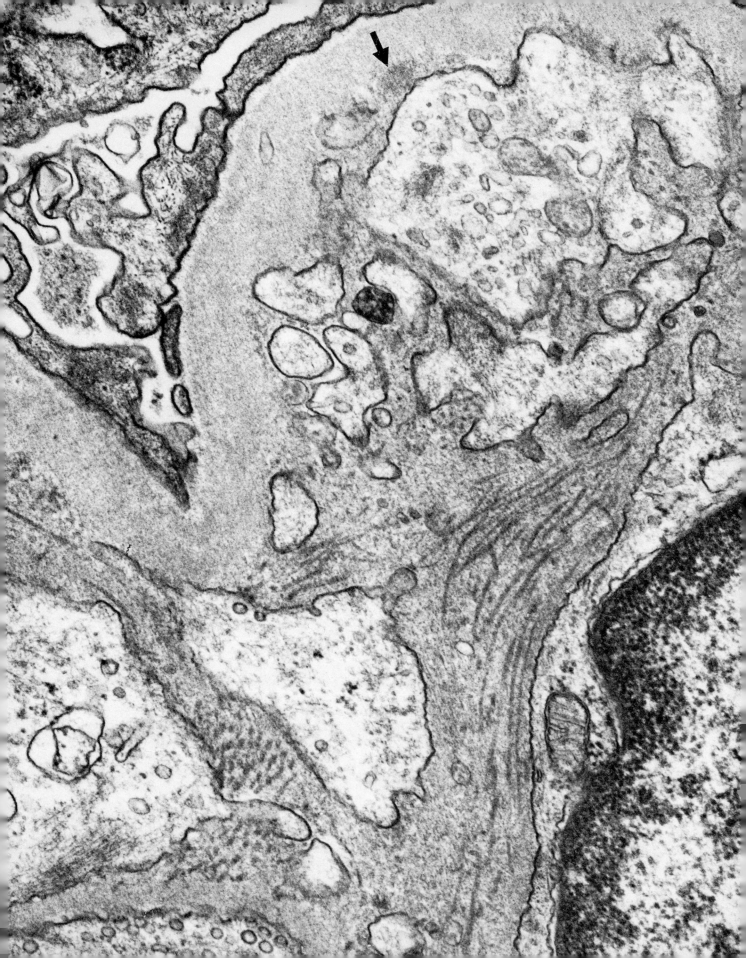

Transplant Glomerulopathy

FIGURE 119. This is a portion of a glomerulus from a human renal allograft 2 years after transplantation. Note the thickening of the basement membrane and the increase in mesangium. There is diffuse rarefaction of the subendothelial side of the basement membrane (arrows) with deposition of fine fibrillar material resembling fibrin aggregates. Similar fibrillar material in greater amounts is also present in the mesangial matrix. The etiology of the gradual deterioration of renal function in long-surviving allograft is not understood. (18,500×)

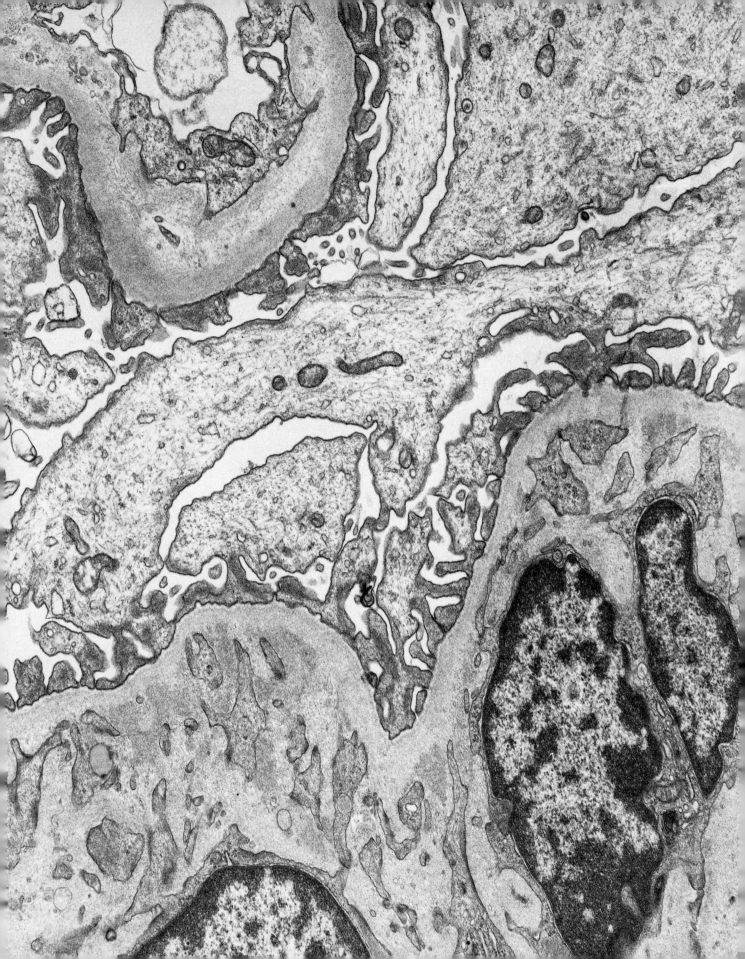

Transplant Glomerulopathy

FIGURE 120. The subendothelial area of the glomerular basement membrane is greatly expanded and contains flocculent granular material and fragments of mesangial cell cytoplasm. (12,700×)

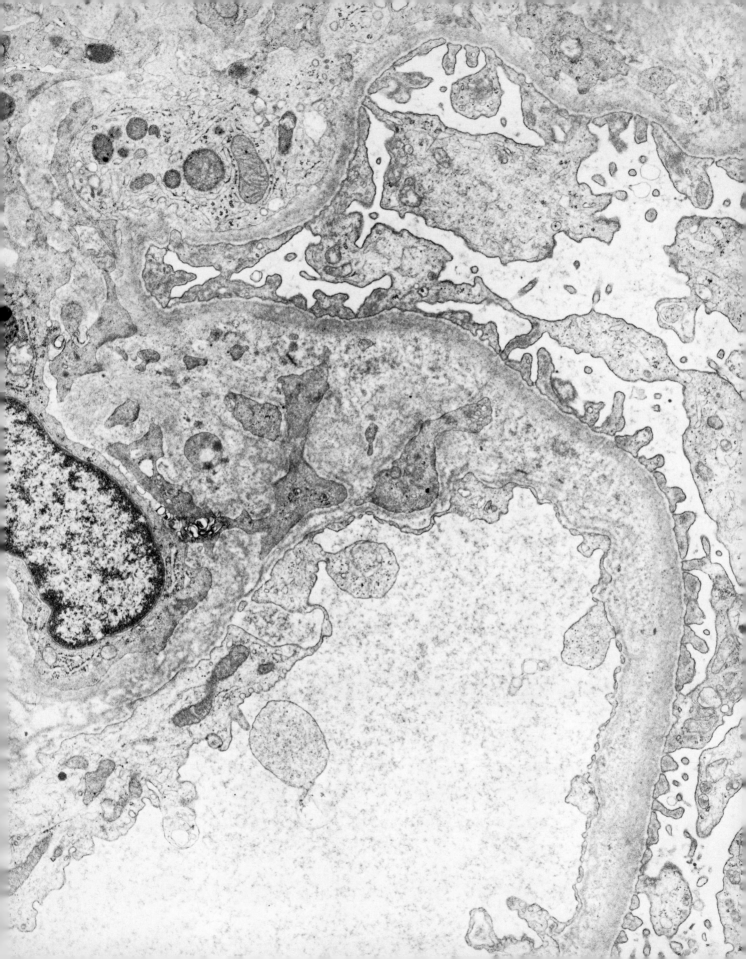

Juxtaglomerular Hyperplasia in the Case of Perimedial Fibroplasia of the Renal Artery

FIGURE 121. An epithelioid cell believed to be a transformed smooth muscle cell is located in the juxtaglomerular apparatus and shows prominent nonspecific granules (lipofucsin-like bodies) and specific granule. The specific granules are characterized by being membrane bound, electron dense, and with occasional crystalloid material in their center. The specific granules are known to contain renin which affects blood pressure. (17,200×)

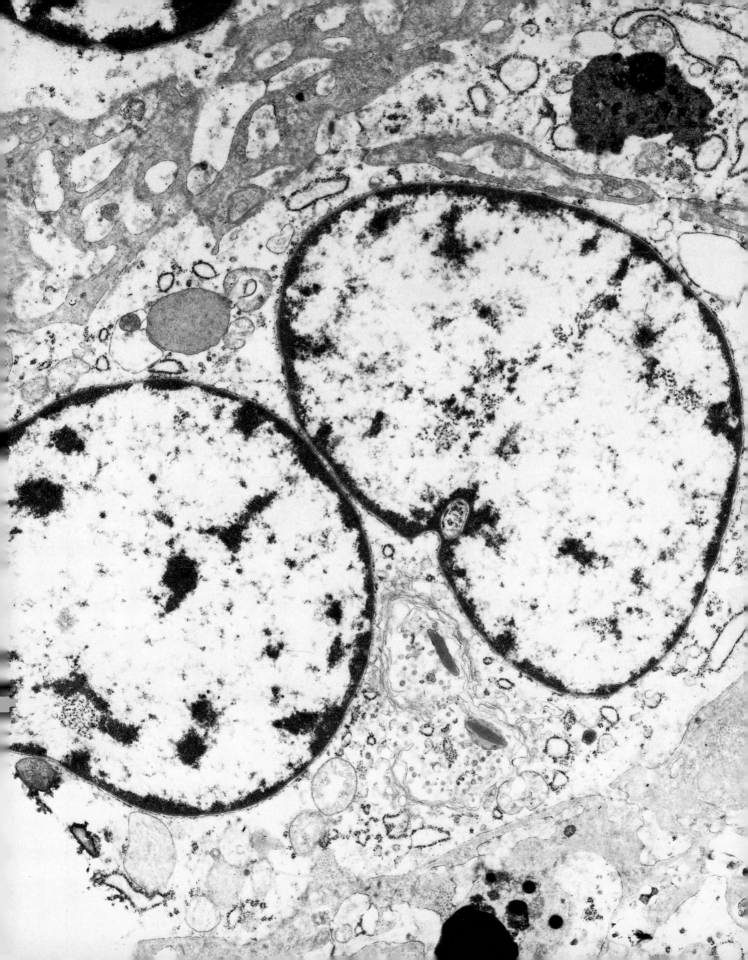

Amyloidosis in the Renal Glomerulus

FIGURE 122. An accumulation of amyloid fibrils in a felt-like array extends across the bottom of the micrograph between a glomerular epithelial cell (closed arrow) and an endothelial cell (open arrow). The pair of smaller arrows within the accumulation of amyloid delineates the basal lamina. At the upper left of the micrograph is a normal arrangement of basal lamina closely sandwiched between epithelial (closed arrow) and endothelial (open arrow) cells and without any amyloid accumulation. A higher magnification of the amyloid fibrils is presented in the next figure. (22,000×)

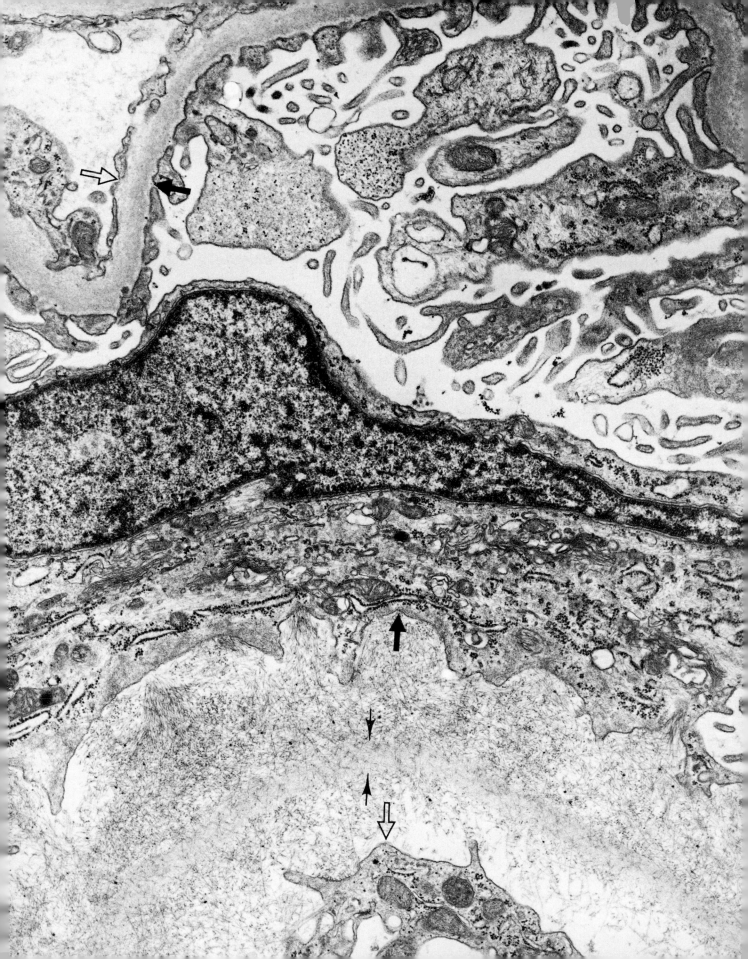

Amyloid Fibrils (High Magnification)

FIGURE 123. Individual amyloid fibrils are haphazardly arranged in a felt-like mesh which can be better appreciated in the low magnification of the preceding figure. The fibrils are each approximately 10 nm in thickness which can be adjudged by comparing their thickness to that of the 10 nm unit membrane (arrow) of the attenuated, fenestrated capillary endothelial cell. (115,000×)

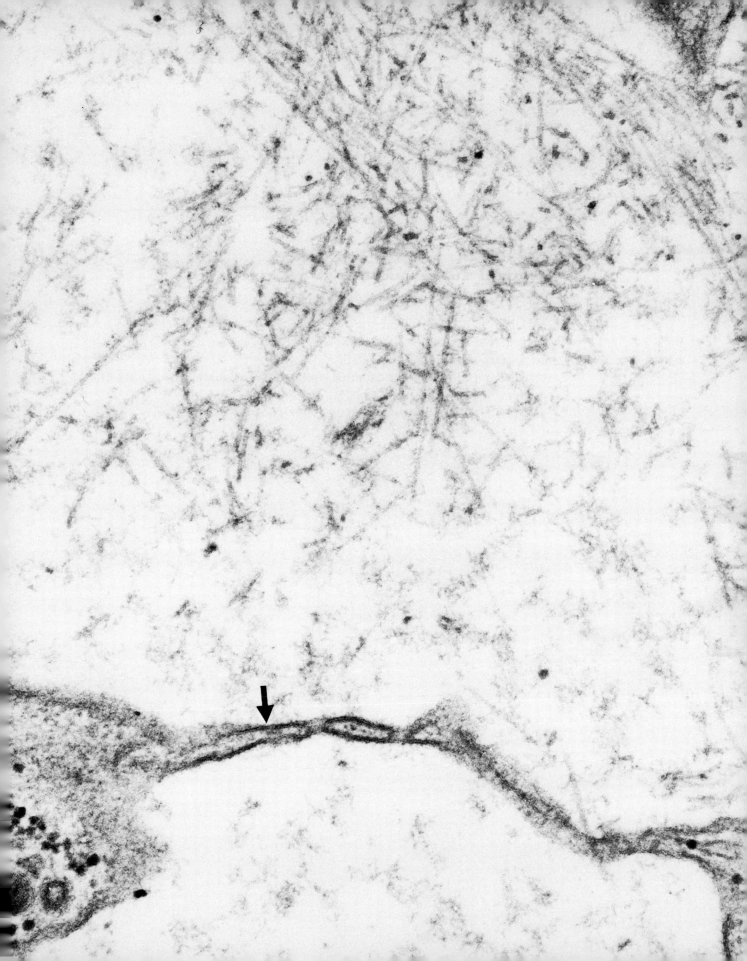

Proximal Tubule Adenoma of the Kidney

FIGURE 124. The oncocytic features of this kidney adenoma are the striking number of mitochondria some of which are slightly swollen and the paucity of most other organelles.

The nuclei are rather rounded and show margination of chromatin. A capillary lies between nests of cells. (13,000×)

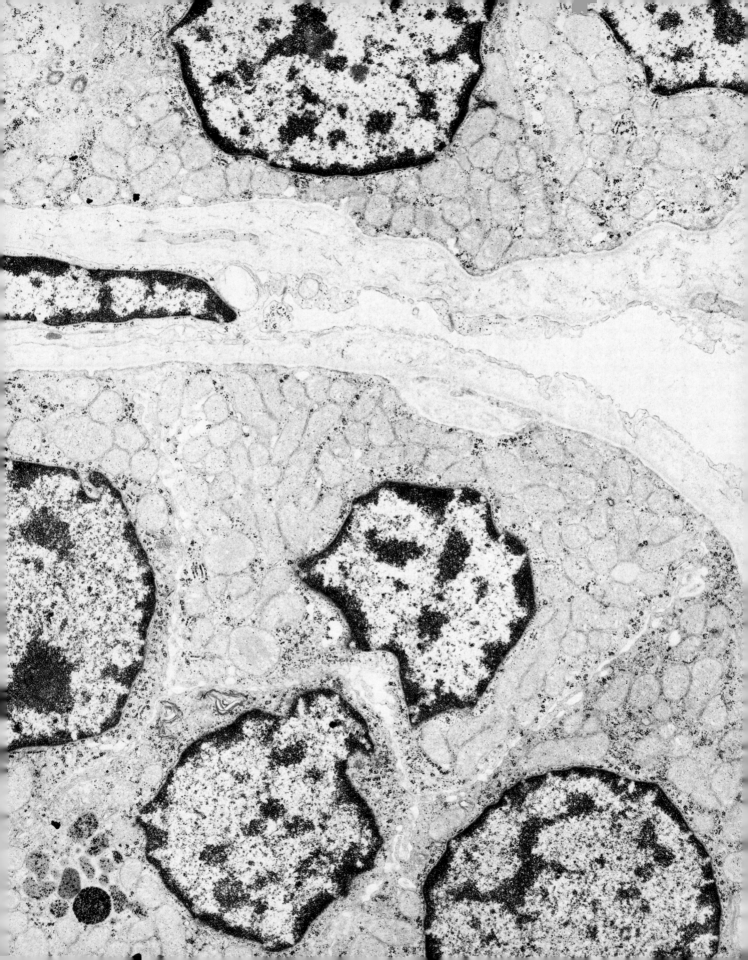

Renal Cell Carcinoma

FIGURE 125. Tumor cells of the clear cell type by light microscopy show by electron microscopy abundant cytoplasm with sparse organelles like mitochondria and lipid. Glycogen granules are numerous and scattered throughout the cytoplasm.

Occasionally there are cells of the granular cell type seen by light microscopy, which by electron microscopy have numerous mitochondria. Interposed between these organelles are glycogen particles (not shown in this picture). It has been speculated that the clear cells are morphologically more like the distal convoluted tubules while the granular cells are more like the proximal convoluted cells. (14,400×)

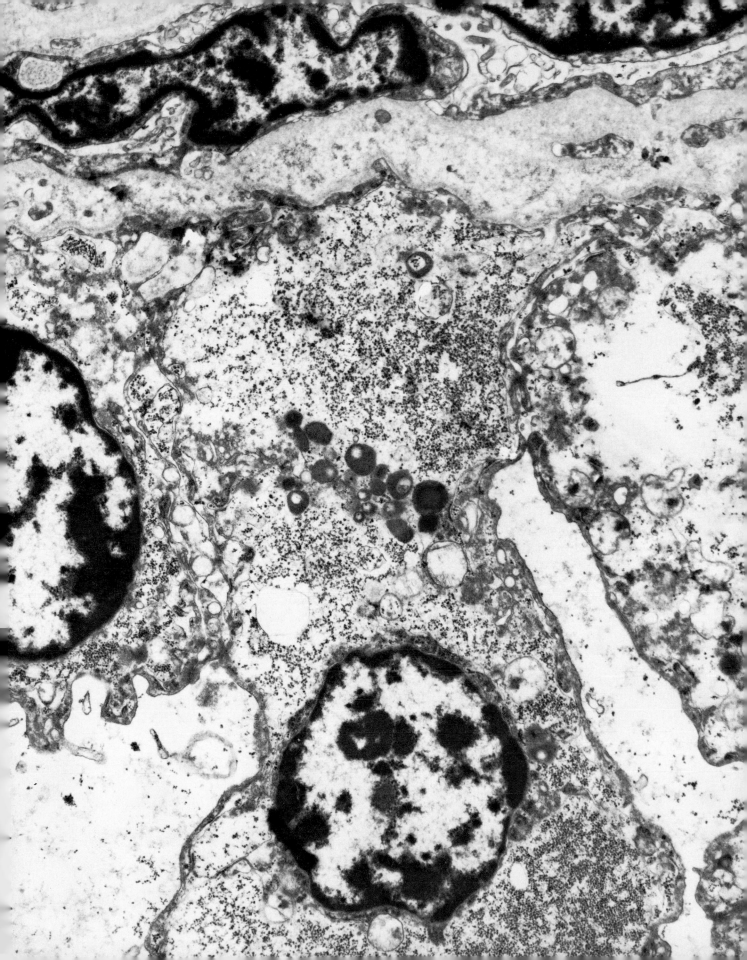

15 Testis

Seminoma

FIGURE 126. This tumor cell is polyhedral in shape and has a large nucleus with a prominent nucleolus. The abundant cytoplasm has few organelles. The cytoplasm is selectively rich in glycogen which aggregates to form rosettes. A small number of polyribosomes are also present. (15,100×)

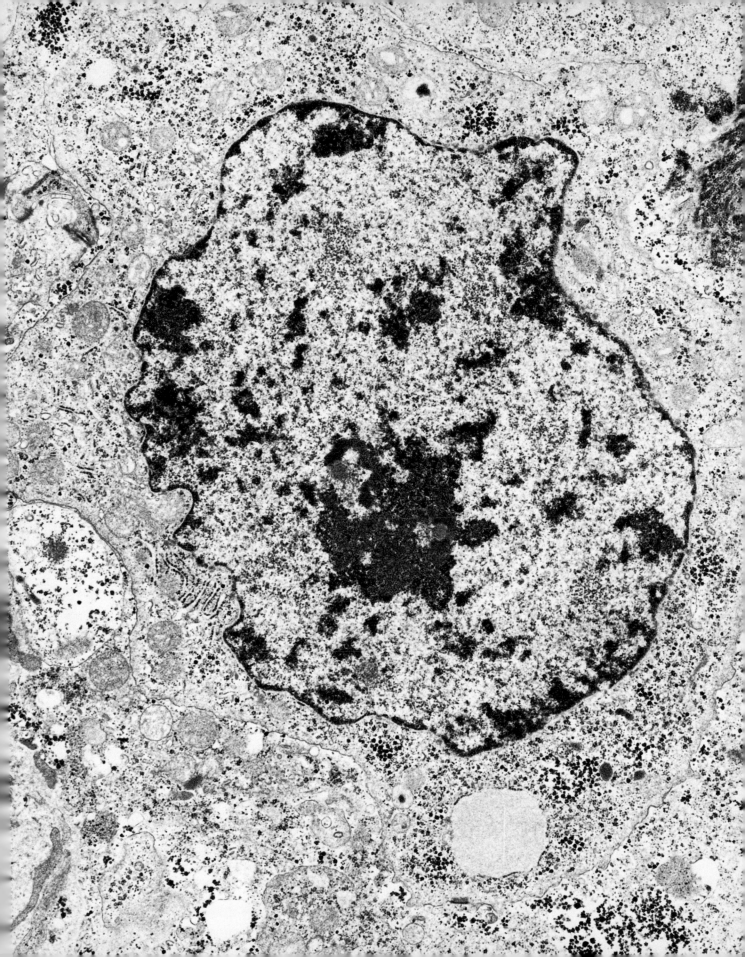

Testicular Embryonal Carcinoma

FIGURE 127. The cell is characterized by a large nucleus with a prominent nucleolus and by scant intracytoplasmic organelles. However, there is abundant glycogen which aggregates to form rosettes. The cell is surrounded by a basement membrane. An adjacent cell shows a large heterogeneous dense body presumably lysosomal in nature. (13,000×)

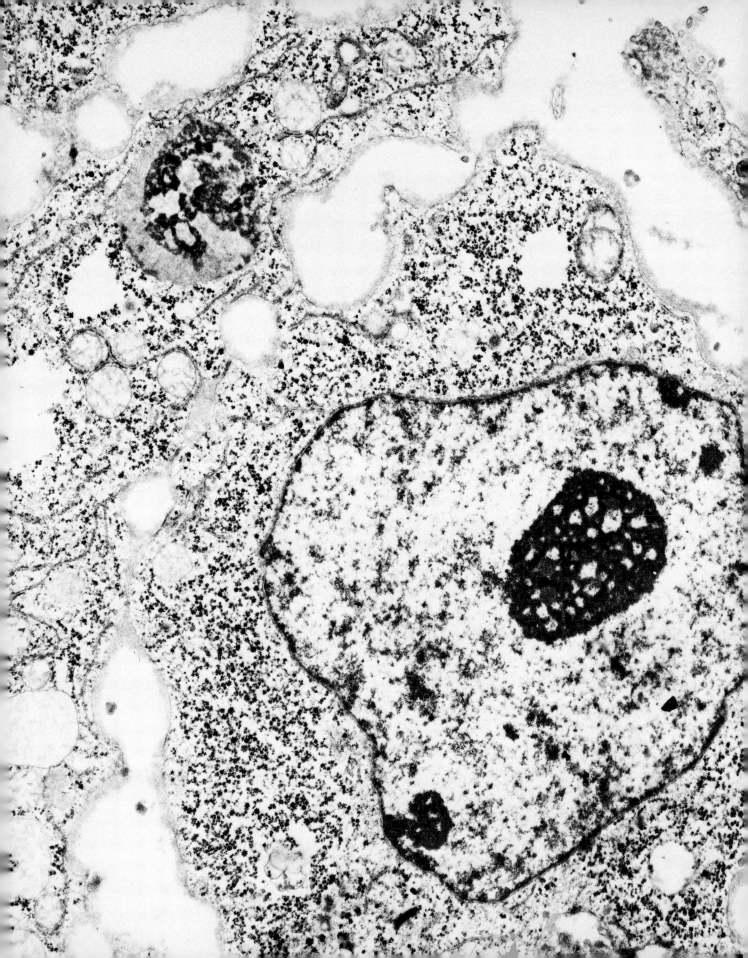

Testis

Testicular Embryonal Carcinoma

FIGURE 128. The tumor cells are closely apposed and are invested by a basement membrane. The cells forming the part of the acinus show short and sparse microvilli. The nuclei are irregular in shape and show prominent nucleoli. The cytoplasmic organelles consist of profiles of RER, a few mitochondria, polyribosomes, and glycogen which aggregate to form rosettes. (11,600×)

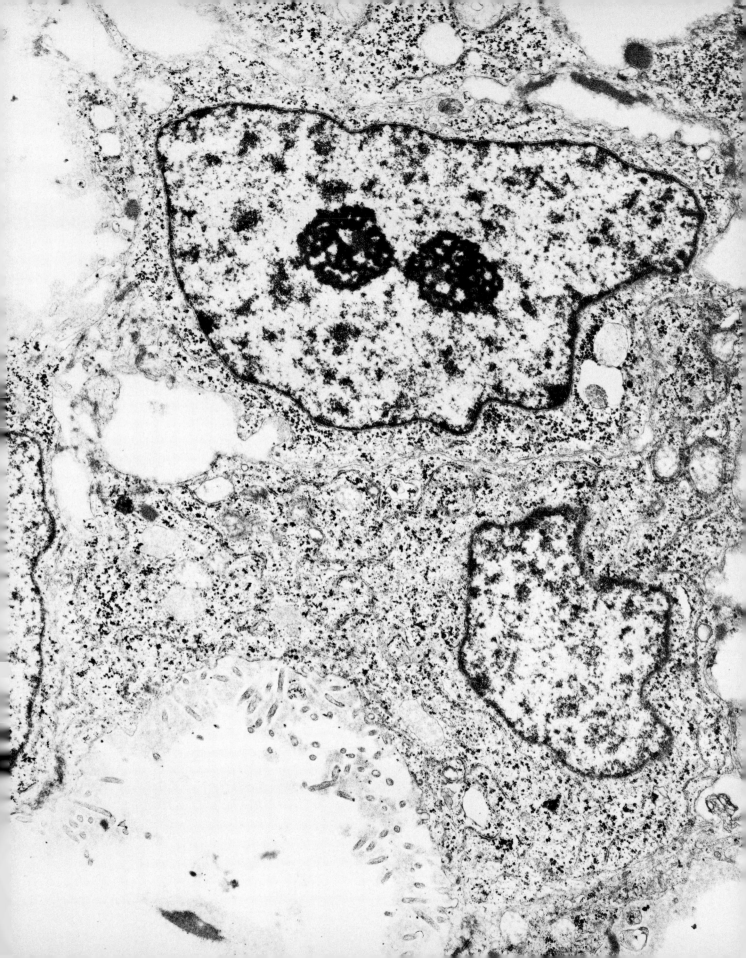

Testicular Embryonal Carcinoma

FIGURE 129. Here in higher magnification, the tumor cells clearly exhibit frequent specialized junctional complexes, the desmosomes. (25,000×)

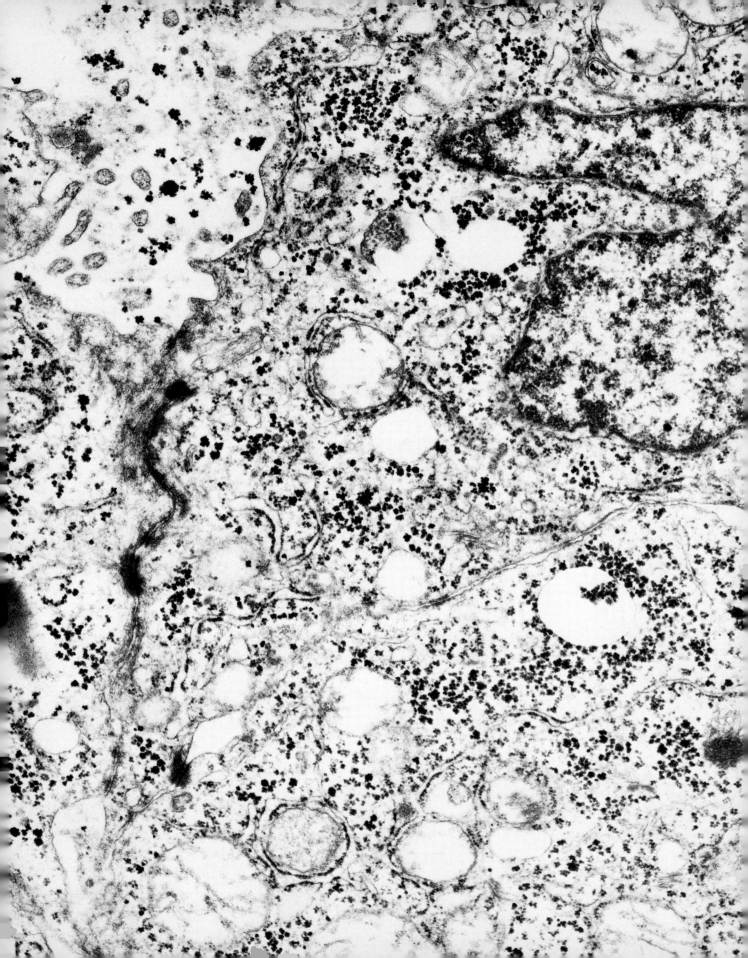

16 Ovary

Brenner's Tumor

FIGURE 130. Cells from the epithelial nests are separated by bundles of collagen and occasionally by basal lamina. Individual cells show a paucity of organelles. Nuclei show indentations and finely dispersed chromatin. (11,000×)

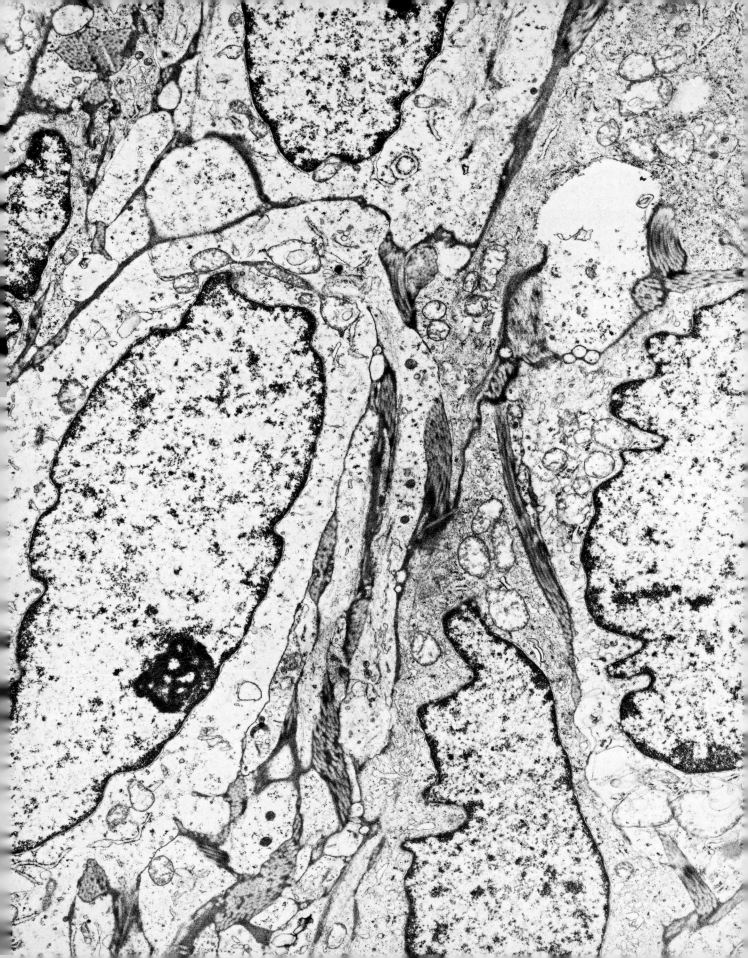

Papillary Serous Cystadenocarcinoma of the Ovary

FIGURE 131. A portion of the papillary configuration is composed of closely apposed columnar cells with interdigitating plasma membrane which are frequently jointed by desmosomes. The luminal surface of the cells possess many short microvilli. The basal part of the cells (lower left in this picture) is decorated by a basal lamina. Electron-dense granules and annulate lamellae are present in the base of the cells. (6000×)

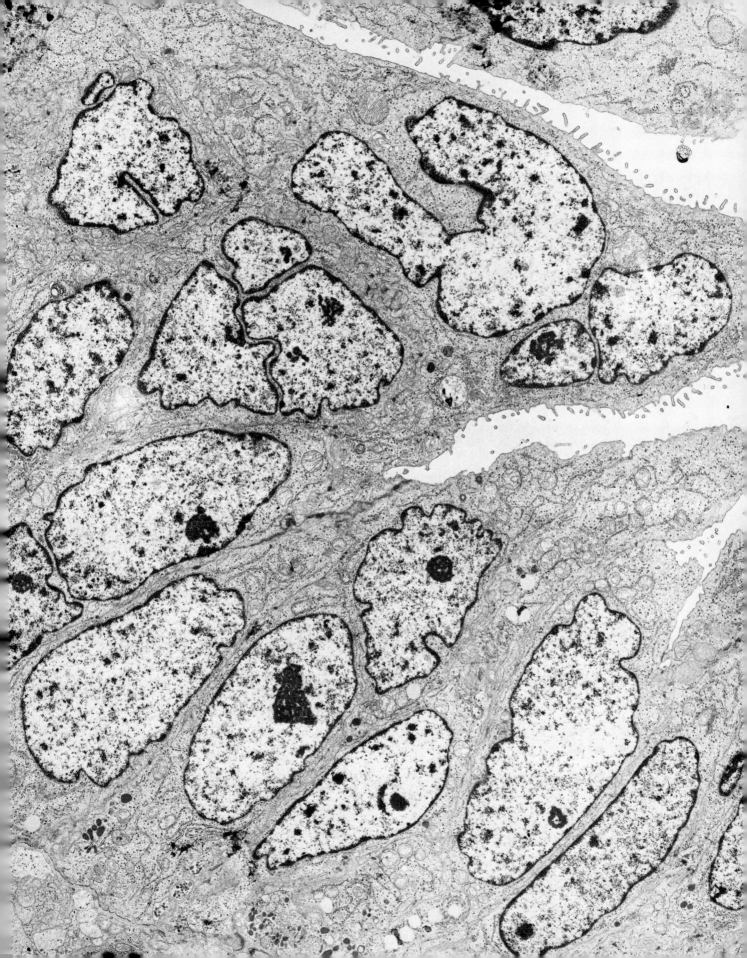

Papillary Serous Cystadenocarcinoma of the Ovary

FIGURE 132. This high magnification of a part of the neoplastic cell shows aggregates of glycogen particles. (22,400×)

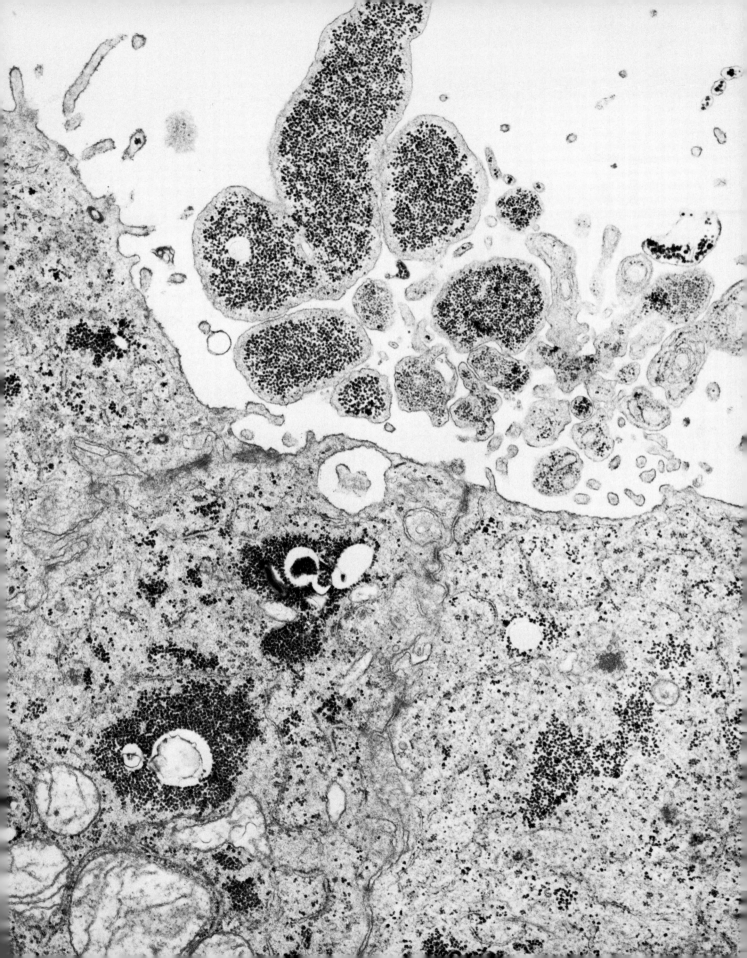

17 Breast

Scirrhous Carcinoma of the Breast

FIGURE 133. A group of carcinoma cells, both epithelial (E) and myo-epithelial (ME) cells is shown. Myoepithelial cells make up the periphery of the cluster and show distinct basement membrane. Adjacent stroma is rich in collagen and elastic fibers. (11,200×)

266

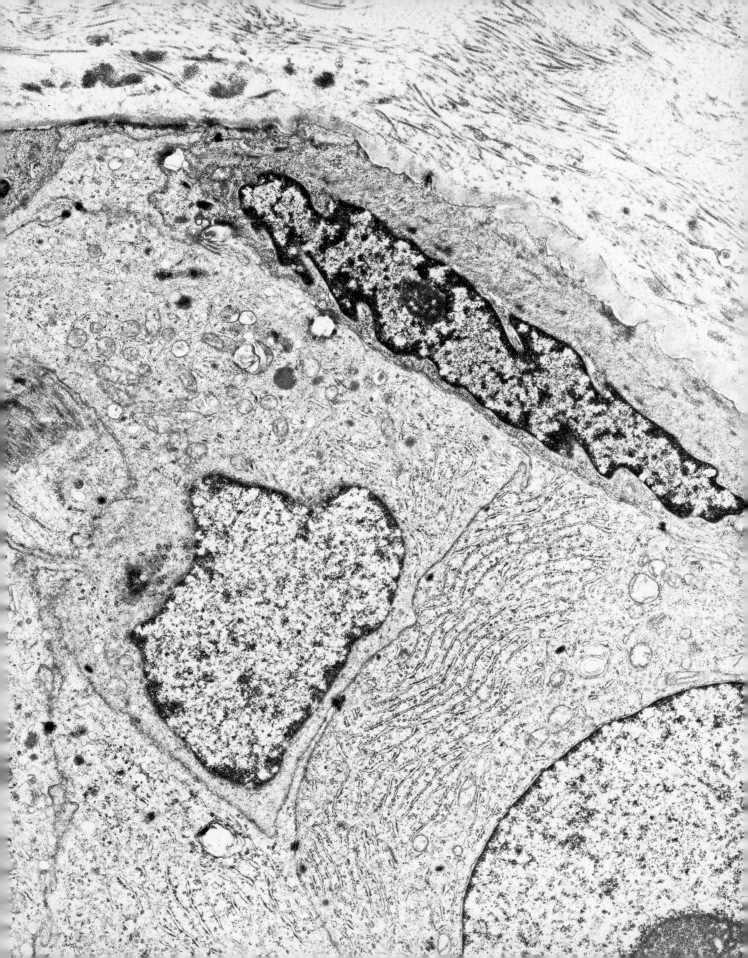

Infiltrating Duct Carcinoma of the Female Breast

FIGURE 134.   This electron micrograph shows clusters of carcinoma cells with round nuclei, and some with nucleoli. There are two types of cells. In the upper left corner there is a cell with dark cytoplasm and many cytoplasmic organelles. The nucleus has an irregular outline and coarse nuclear chromatin. The other cells have round nuclei and paler cytoplasm with fewer cytoplasmic organelles. The lowest cell shows some dense membrane bound secretory granules. The cells are apposed together by desmosomal attachment. (24,500×)

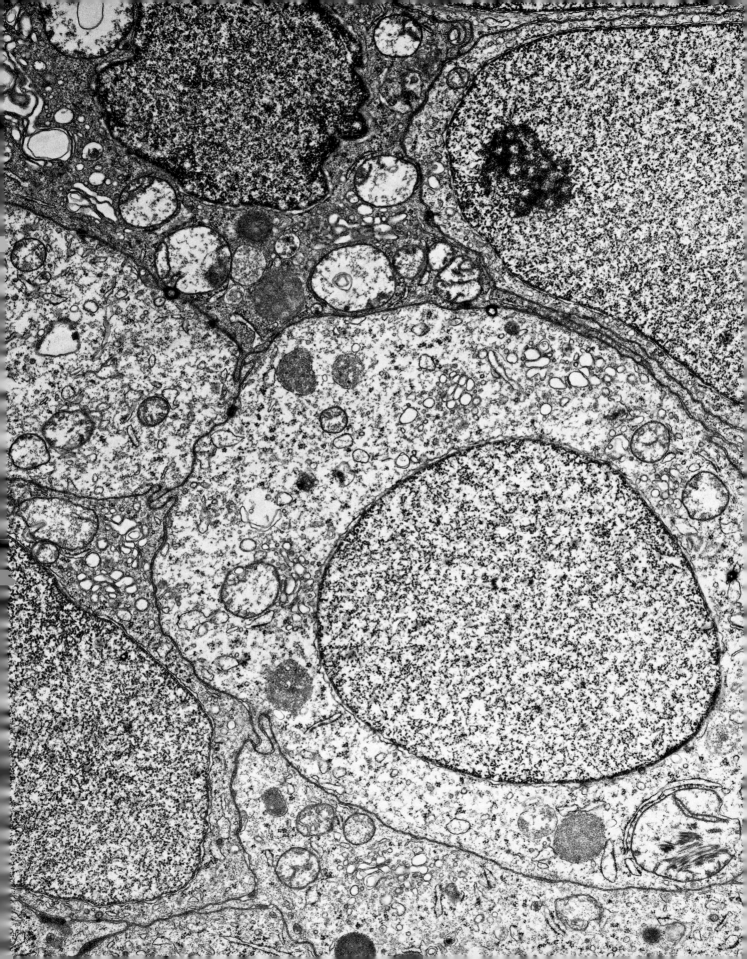

Small Cell Carcinoma of the Male Breast

FIGURE 135. The cells are closely packed usually with round nuclei and prominent nucleoli. Organelles are rather scant and consist of mitochrondria, free ribosomes and rough endoplasmic reticulum. Some cells also contain cytoplasmic fibrils. Adjacent cell membranes are closely apposed with few desmosomes. (8600×)

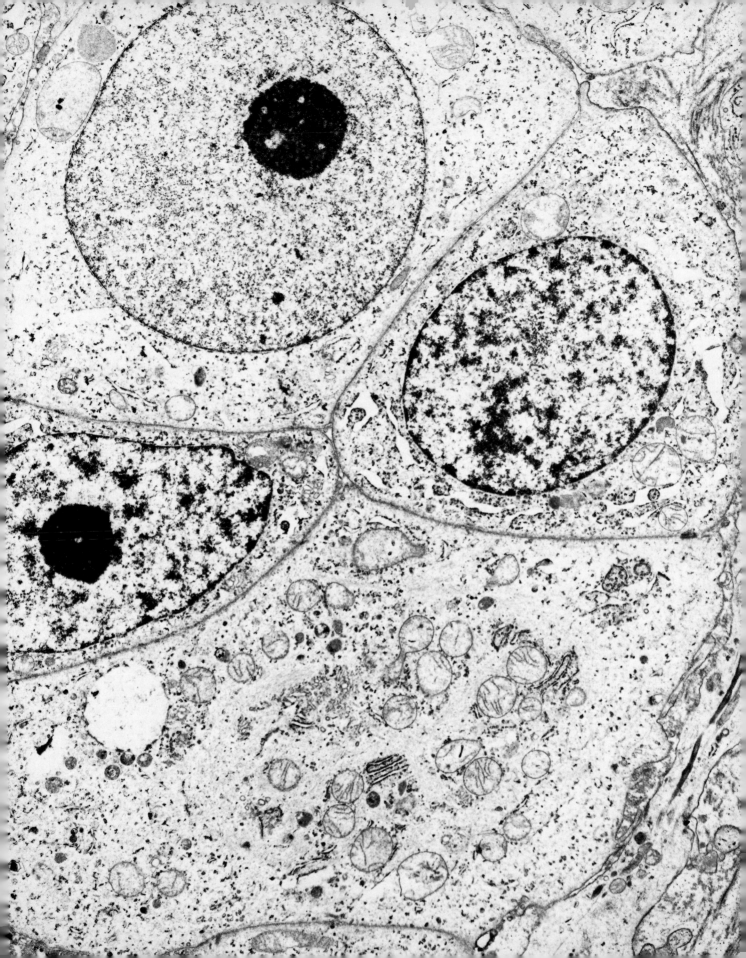

18 Soft Tissue

Normal Skeletal Muscle

FIGURE 136. This longitudinal section of myofibers shows many myofibrils separated by rows of mitochondria in a regular fashion. The cells are enveloped by a plasma membrane, called the sarcolemma, which is coated by a thin layer of an osmiophilic basement membrane. The saccolemma shows multiple small invaginations. The myofibrils are composed of thick filaments (myosin) and thin filaments (actin). (36,800×)

Degeneration of Skeletal Muscle

FIGURE 137. In this portion of a muscle fiber, note areas of central myofibrillar degeneration. In the degenerated areas there is disarry and dissolution of myofibrils accompanied by loss of mitochondria. These changes can be seen in any muscle injury but more commonly in muscular dystrophy, polymyositis, or other severe muscle injuries. (26,200×)

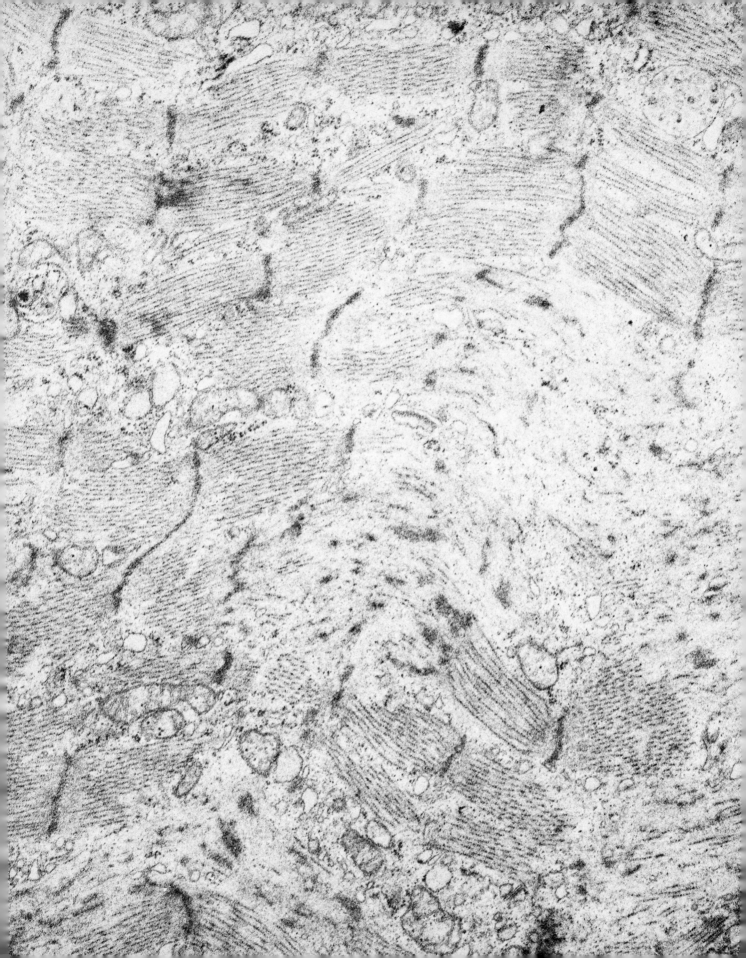

Regeneration of Skeletal Muscle

FIGURE 138.   This micrograph shows a cluster of myoblasts in an area of regeneration. The myoblasts have a large central nucleus and are rich in rough endoplasmic reticulum. The cells contain varying amounts of myofilaments in haphazard arrangement. Pinocytic vesicles are numerous along the plasma membranes. (18,400×)

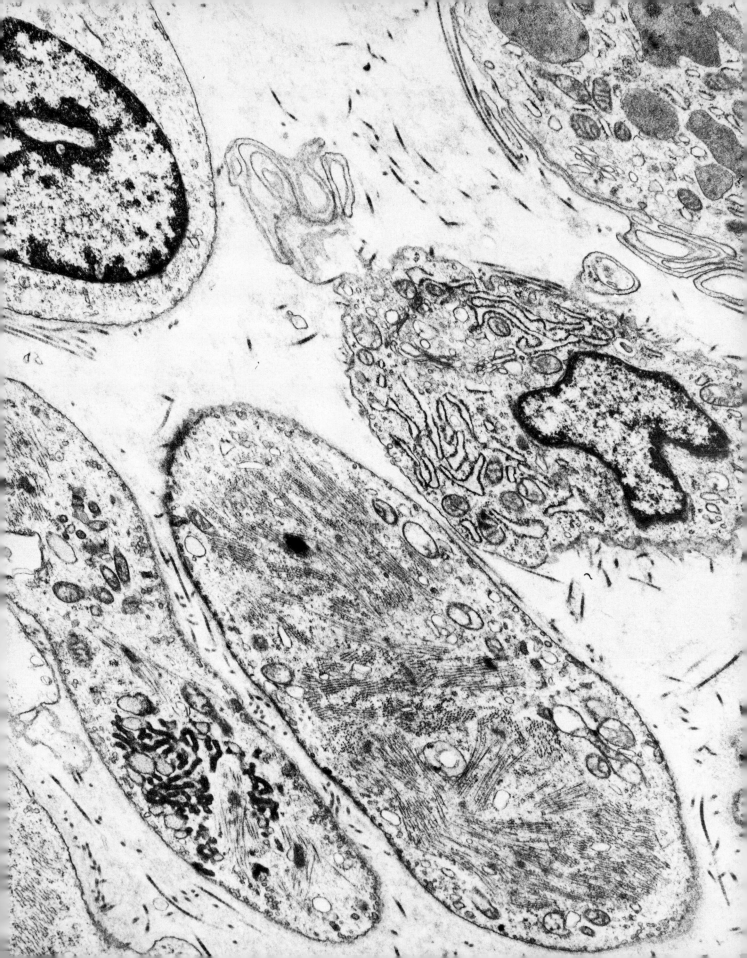

Nemaline Myopathy

FIGURE 139. This micrograph demonstrates two muscle fibers from a youngster who presented with mild muscle weakness since birth. Note the many rod-like structures in both fibers. They appear to arise from Z-bands. This is one of the better known forms of congenital myopathy with distinct morphologic features. Some of the reported cases suggested an autosomal dominant form of inheritance. (21,500×)

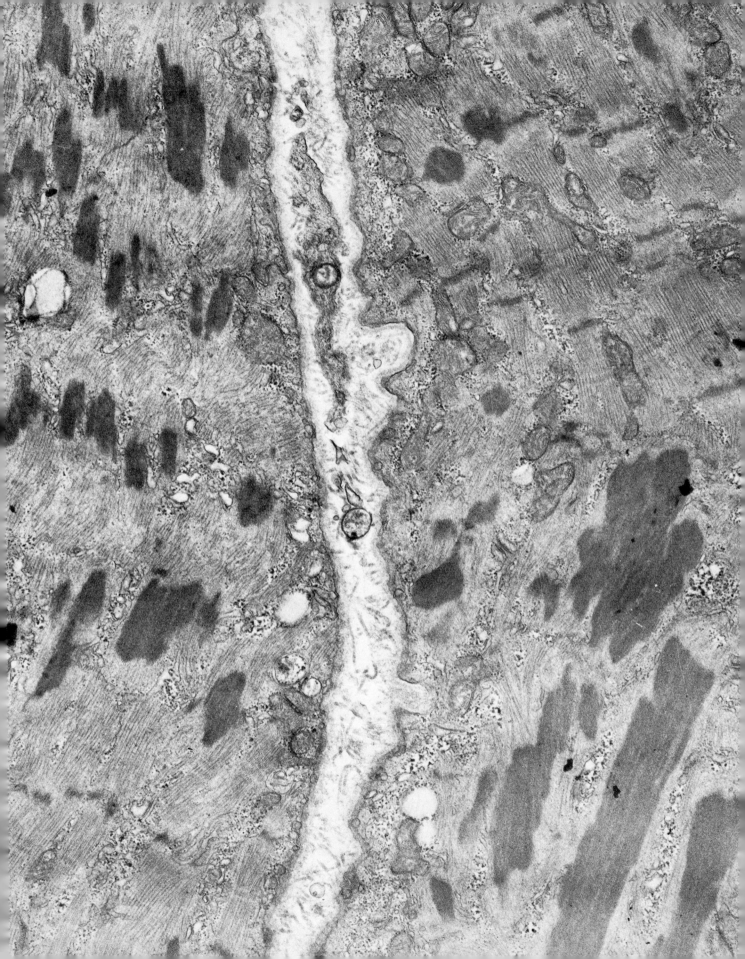

Nemaline Myopathy

FIGURE 140.   This is a high magnification of the rod-like structures. Note the longitudinal periodicity in these bodies similar to that of the Z-band. (63,800×)

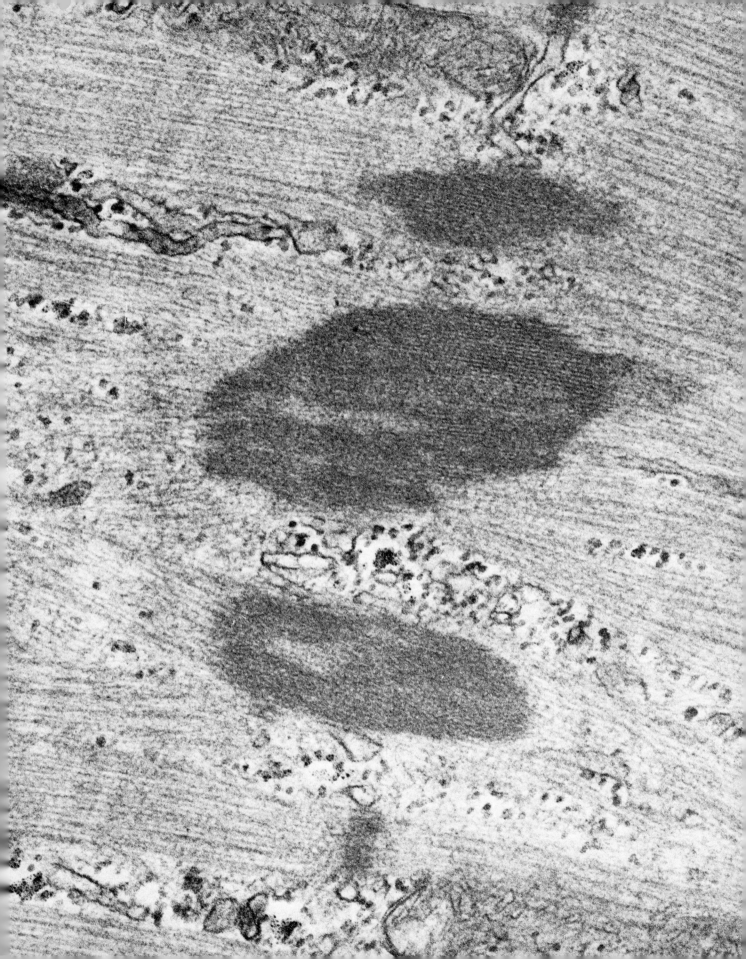

Mitochondrial Myopathy

FIGURE 141. Cross sections of muscle fibers from a young boy with mild relatively nonprogressive muscle weakness are shown. Note the tremendous proliferation of mitochondria in the subsarcolemmal and the intermyofibrillar spaces of the muscle fibers. There is a marked increase in cristae in each mitochondria, and they occasionally contain paracrystalline structures. A myelinated nerve fiber is seen in the interstitium. Mitochondrial myopathy is one of the most common forms of congenital myopathy. It is heterogenous in origin and has been known to be associated with various metabolic diseases. (15,900×)

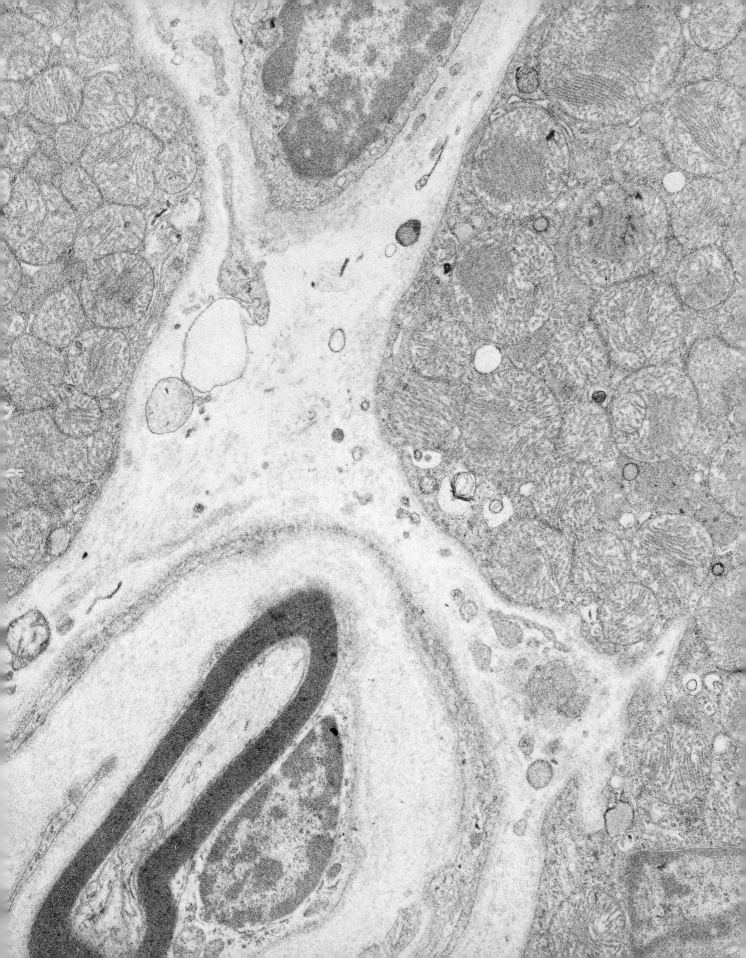

McArdle's Syndrome

FIGURE 142. Electron microscopy reveals excess glycogen deposited within the striated muscle and degeneration of the mitochondria. (21,000×)

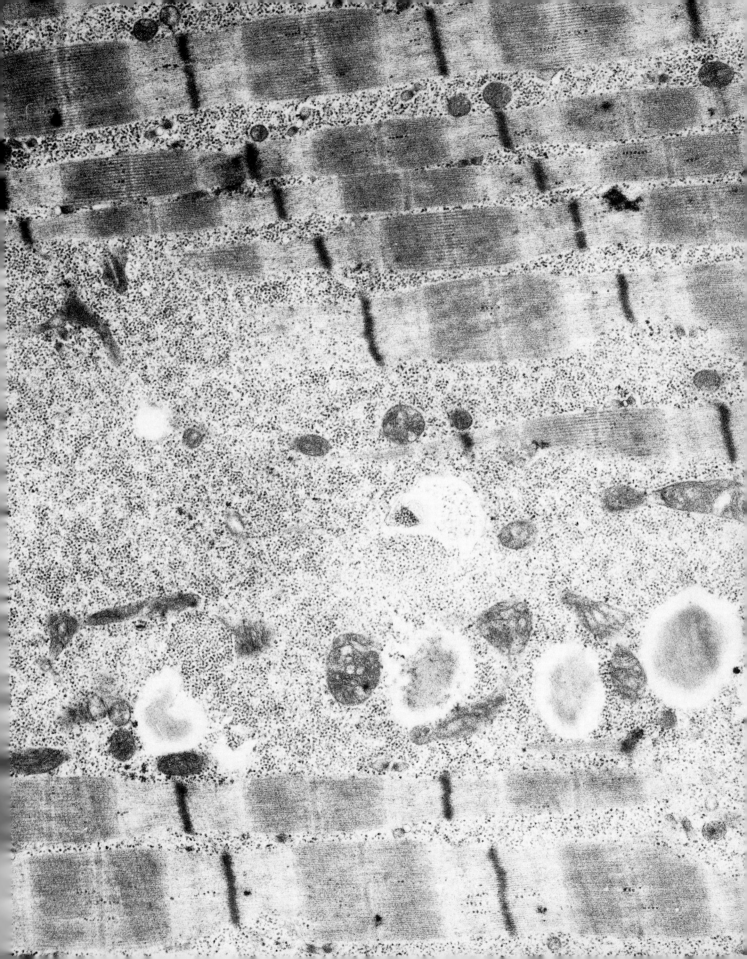

Skeletal Muscle in Glycogenosis, Type 11 (Pompe's Disease)

FIGURE 143. This is a portion of a skeletal muscle fiber from a 3-month-old infant with diffuse cardiac enlargement and severe hypotonia. Note that the muscle fiber is distended with rosette type glycogen particles, some of which are bounded by a limiting membrane. A small aggregate of myofilaments with a Z-band, representing the residual myofibril is seen between two glycogen-filled cytosomes. Pompe's disease is a lysosomal disease with a deficiency of acid maltase. Although the disease can affect many organs, functional impairment is most striking in heart and skeletal muscle. (49,400×)

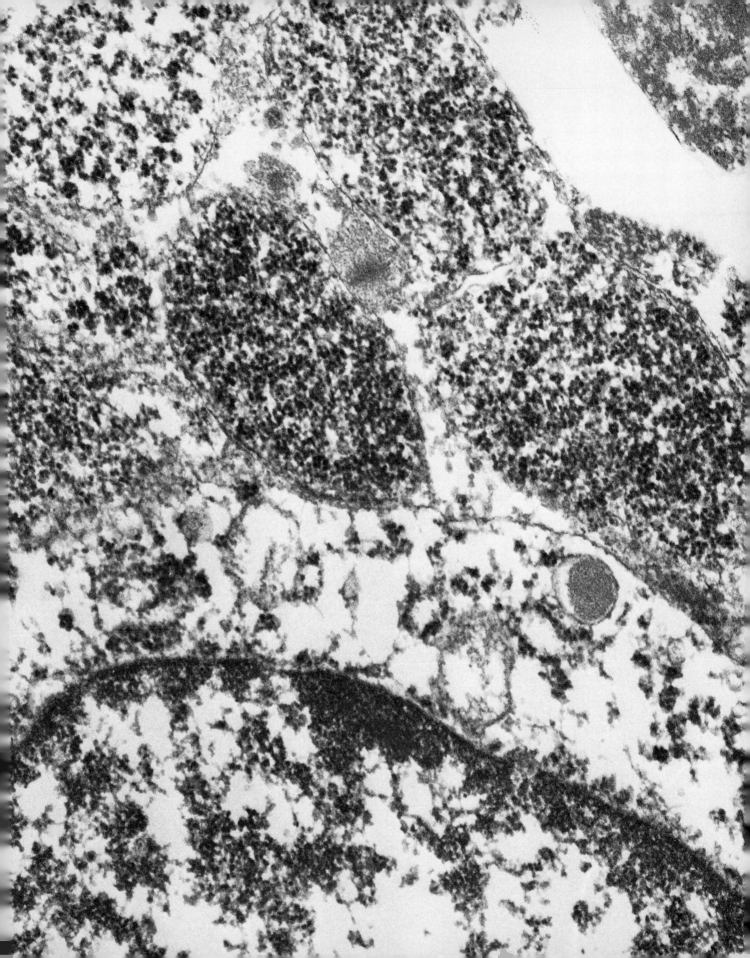

Ceriod Lipofuscinosis, Late Infantile Type

FIGURE 144. A part of a smooth muscle cell (pericyte) from skin biopsy is shown. Note several unusual cytosomes (arrows) in the cytoplasm. The cytosomes have a single outer membrane and are filled with curvilinear profiles (CCP). The structures are practically diagnostic for ceriod lipofuscinosis (Batten's disease). They are found in many different kinds of cells but skin or muscle is considered the material of choice. (mf: myofibrils, 38,800✕)

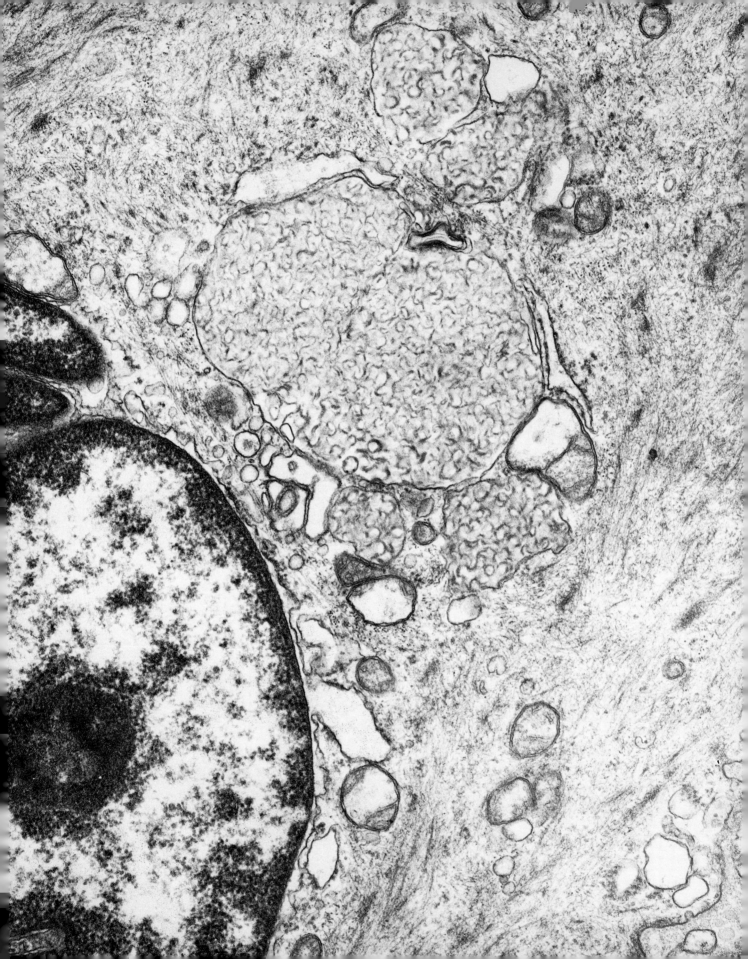

Granular Cell Tumor

FIGURE 145. The characteristic features in cells from granular tumors are cytoplasmic granules. The granules are heterogenous, are membrane-bound, contain electron-dense material, and are thought to be lysosomal in nature. Another notable feature is the basal lamina, multi-layered in this case, seen at the basal aspect of the cell. (31,000×)

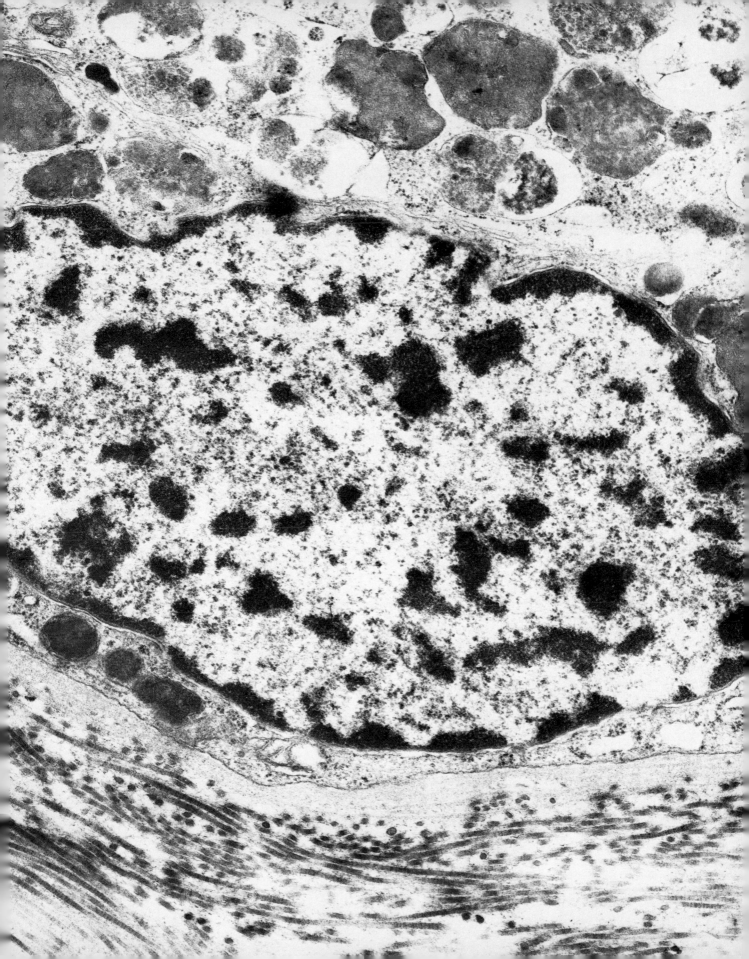

Alveolar Soft Part Sarcoma

FIGURE 146. The cells in alveolar soft part sarcoma contain cytoplasmic crystals of varying sizes, shapes, and density. The majority of these crystals are surrounded by a limiting membrane. These crystals are composed of white and dark lines with a regular periodicity. (25,000×)

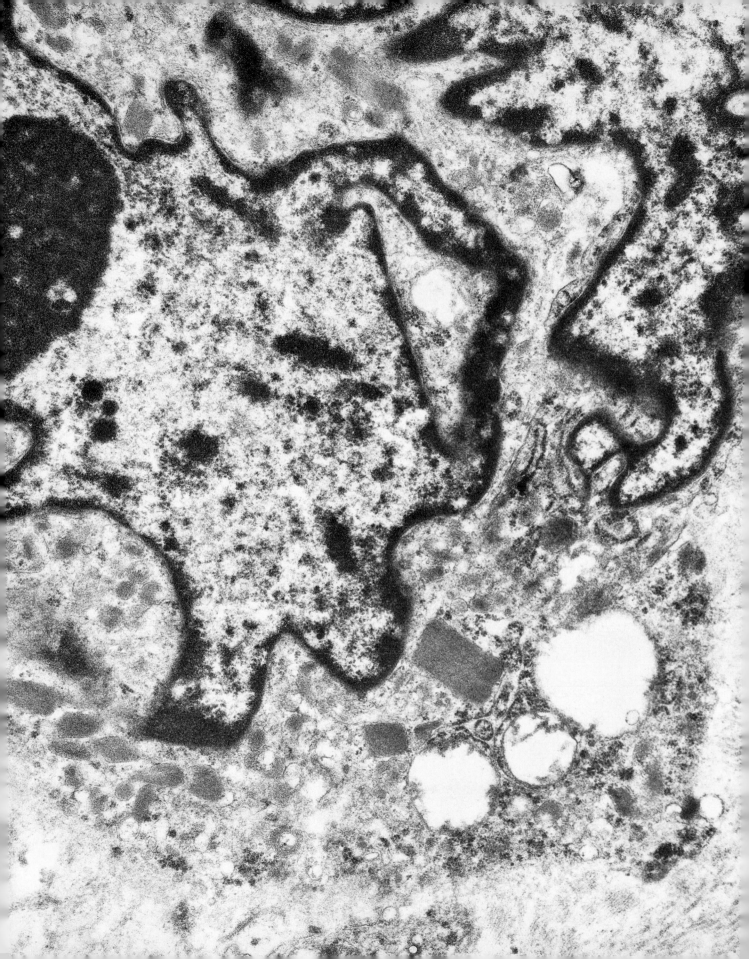

Hemangiopericytoma

FIGURE 147.   A small capillary is surrounded by spindle-shaped cells. The capillary is partially invested by its basal lamina. The surrounding spindle cells show characteristic features of the basal lamina, pinocytotic vesicles and occasional microfilaments. (10,900×)

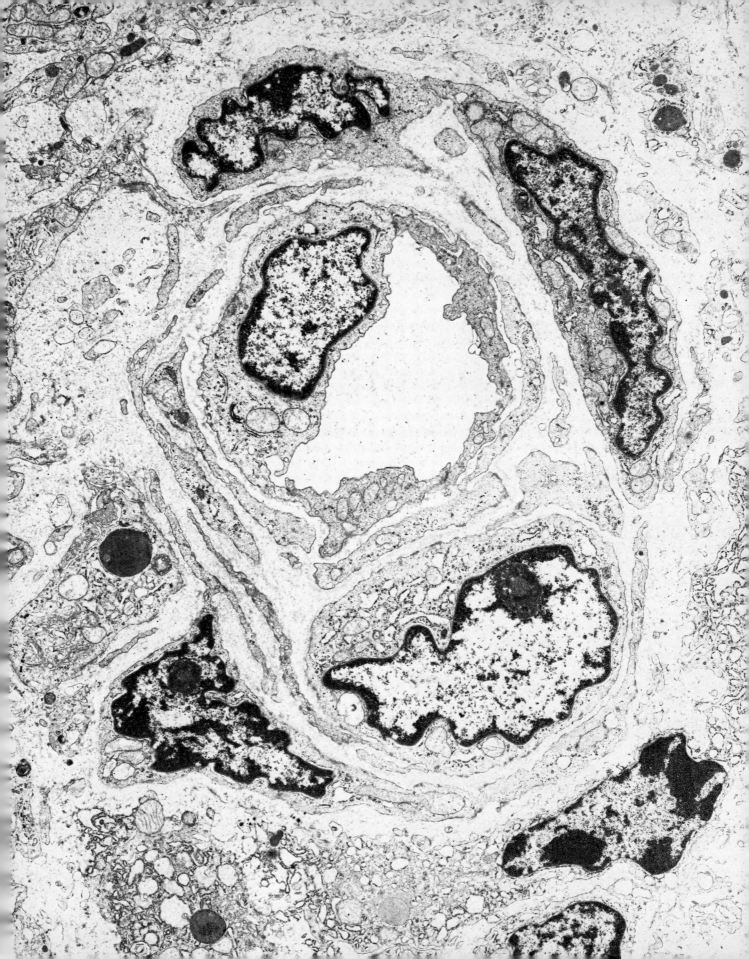

Hemangiopericytoma

FIGURE 148.   The tumor cells are closely apposed, spindle shaped, and have a basal lamina. Fine filaments with occasional condensation into small fascicles are a striking feature of the tumor cell cytoplasm. The condensations are mostly distributed along the inner aspect of the plasmalemma. The plasma membrane is commonly provided with many pinocytotic vesicles and exhibits small inconspicuous zones of attachment, reminiscent of weak or half-desmosomes. The basal lamina, cytoplasmic filaments and pinocytotic vesicles are characteristic findings for hemangiopericytoma. (10,900×)

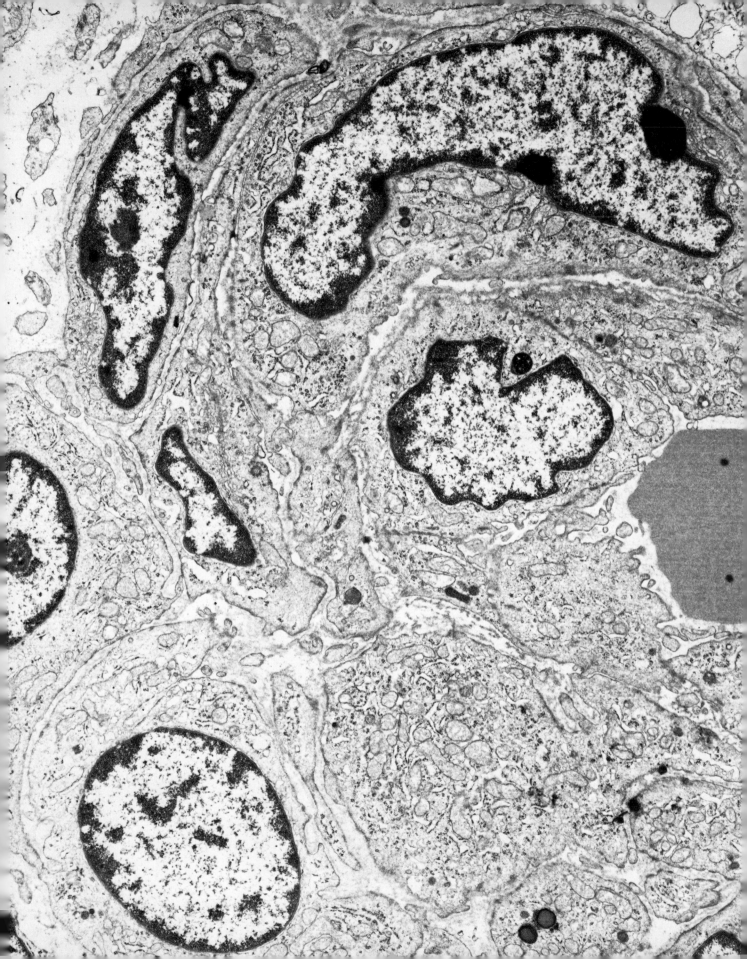

**Malignant Fibrous Histiocytoma—
Multinucleated Giant Cell**

FIGURE 149. The multinucleated giant cell shows finger-like cytoplasmic projections, more than two nuclei, abundant rough endoplasmic reticulum, numerous mitochondria, occasionally intranuclear bodies, microfilaments, and lysosomes. (9200×)

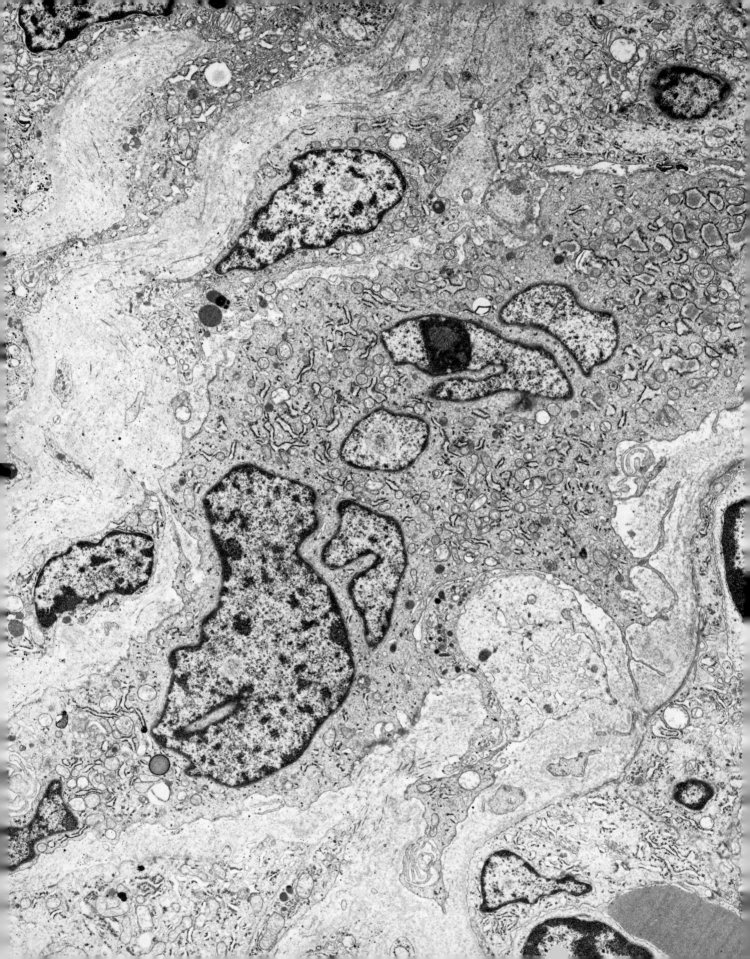

Malignant Fibrous Histiocytoma—Fibroblast-like Cell

FIGURE 150.   The fibroblast-like cell has an elongated nucleus with nuclear invaginations. The cytoplasm is rich in rough endoplasmic reticulum and also contains many mitochondria and one or two well-developed Golgi zones. (14,700×)

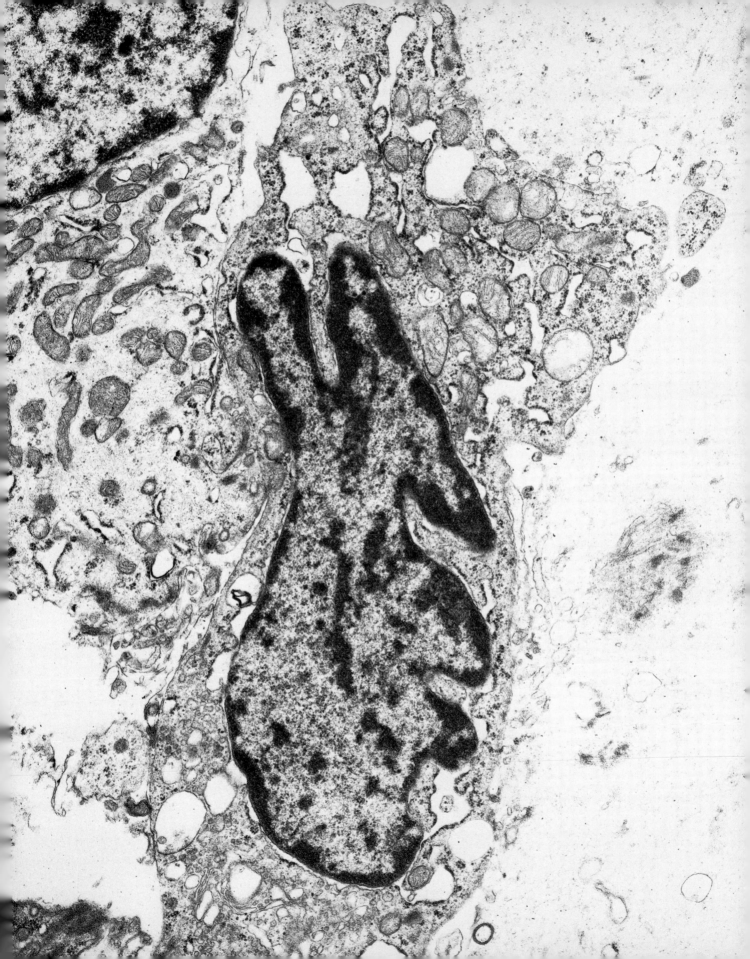

Malignant Fibrous Histiocytoma—Xanthomatous Cell

FIGURE 151. Xanthomatous cell contains numerous lipid droplets, which displace the nucleus to one side. (17,800×)

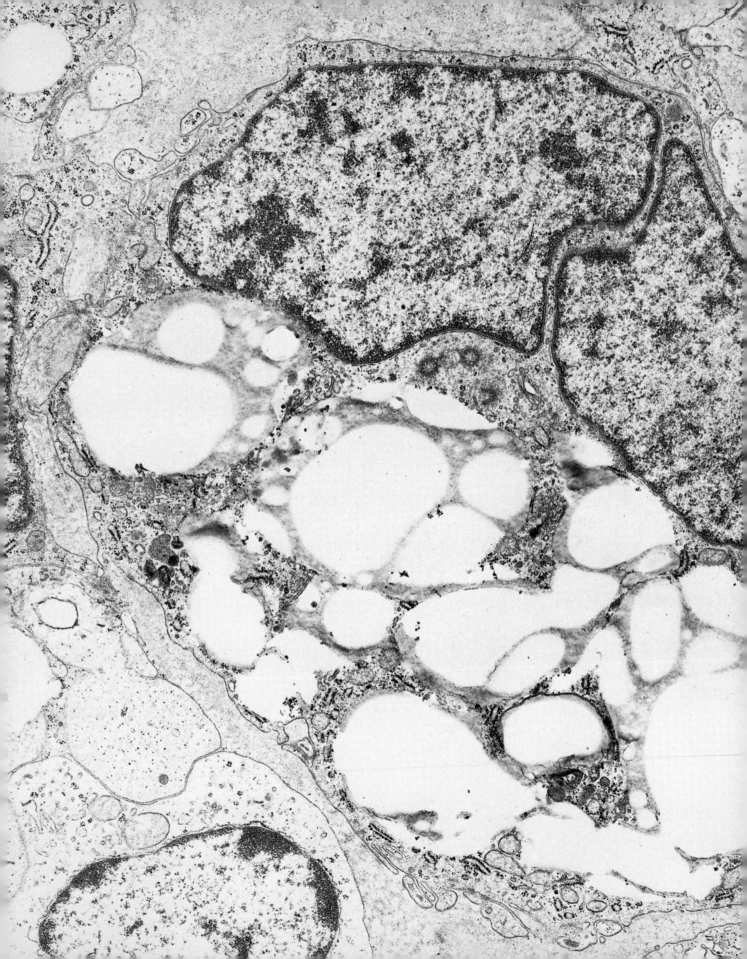

Rhabdomyosarcoma

FIGURE 152. The neoplastic cell running diagonally across this electron micrograph contains the essential hallmark of a striated muscle cell. Thick and thin myofilaments are present, and they form the classical banding of striated muscle albeit not as well organized. This cell is undoubtedly part of a rhabdomyosarcoma. (7600×)

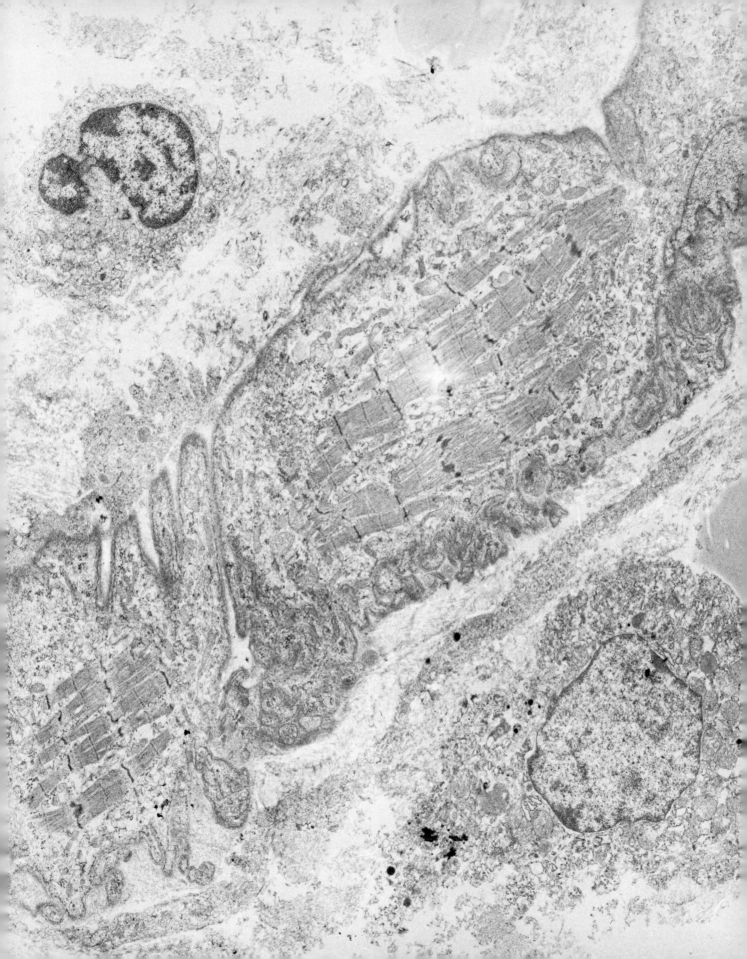

Liposarcoma

FIGURE 153. The electron micrograph shows clusters of cells with cytoplasm rich in vacuoles. The nuclei show dense heterochromatin mostly arranged around the nuclear membrane. (13,000×)

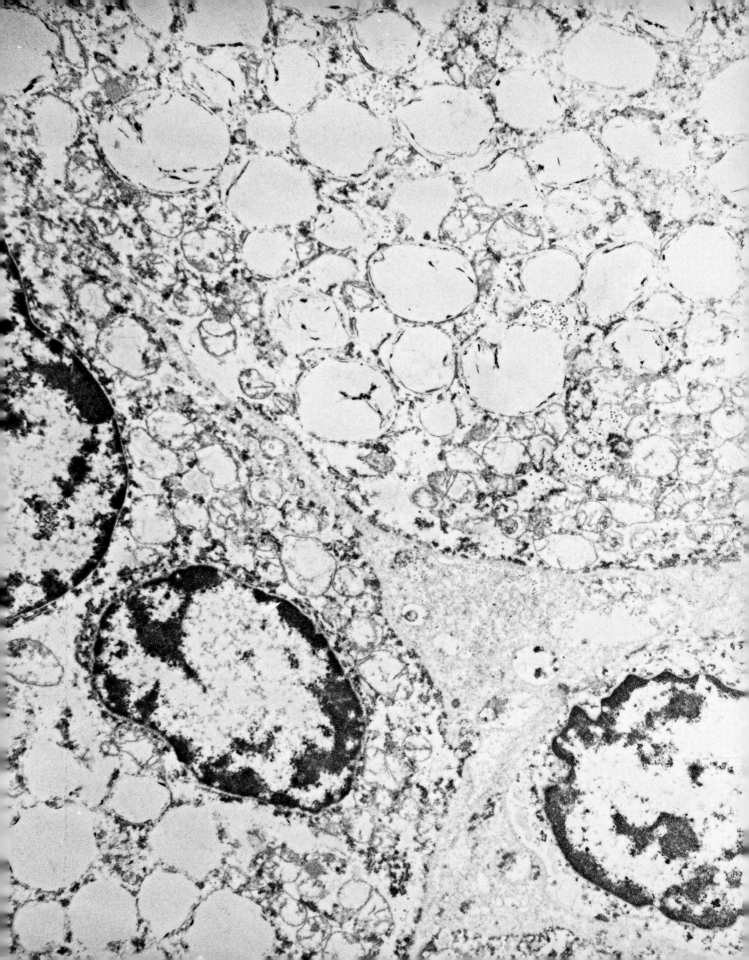

19 Bone and Joint

Osteosarcoma

FIGURE 154.   This electron micrograph presents a portion of the cytoplasm and nucleus of a malignant osteoblast. The typical malignant osteoblast, like its benign counterpart, displays a well-developed Golgi complex and rough endoplasmic reticulum. The cytoplasm of the tumor cell is rich in RER which is much dilated forming intercommunicating irregular cavernous spaces containing flocculent material. The Golgi system is made of small vesicles and membranous structures occupying large areas of the perinuclear cytoplasm. In addition, mitochondria, lysosomes, glycogen particles, ribosomes, polyribosomes, and filaments are seen throughout the cytoplasm. The adjacent intercellular space shows a low electrodense matrix enclosing haphazardly arranged collagen and elastic fibers. Other neoplastic cells that are also rich in RER are those of chondrosarcoma and fibrosarcoma. (21,600×)

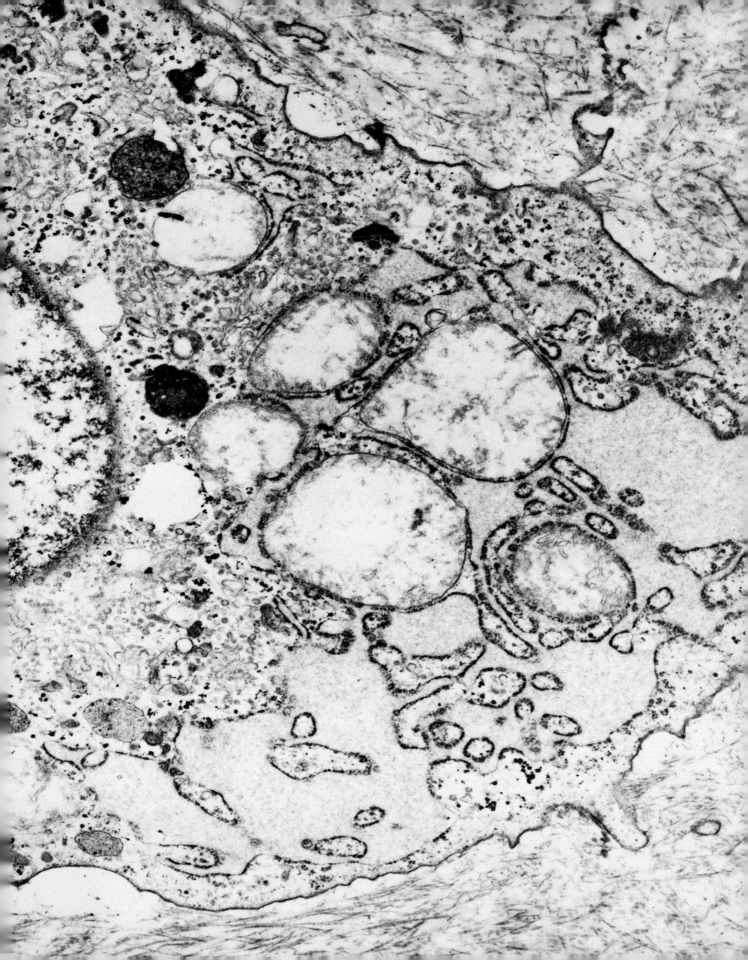

Osteosarcoma

FIGURE 155. A poorly differentiated neoplastic osteoblast from an osteosarcoma displays scant cytoplasm with a few short profiles of RER, polyribosomes, ribosomes, vesicles containing flocculent material and a dense body. Intracytoplasmic filaments are observed along the cell membrane which is scalloped. The nucleus contains euchromatin and the nuclear membrane is slightly uneven. The extracellular space is rich in collagen fibers. (25,100×)

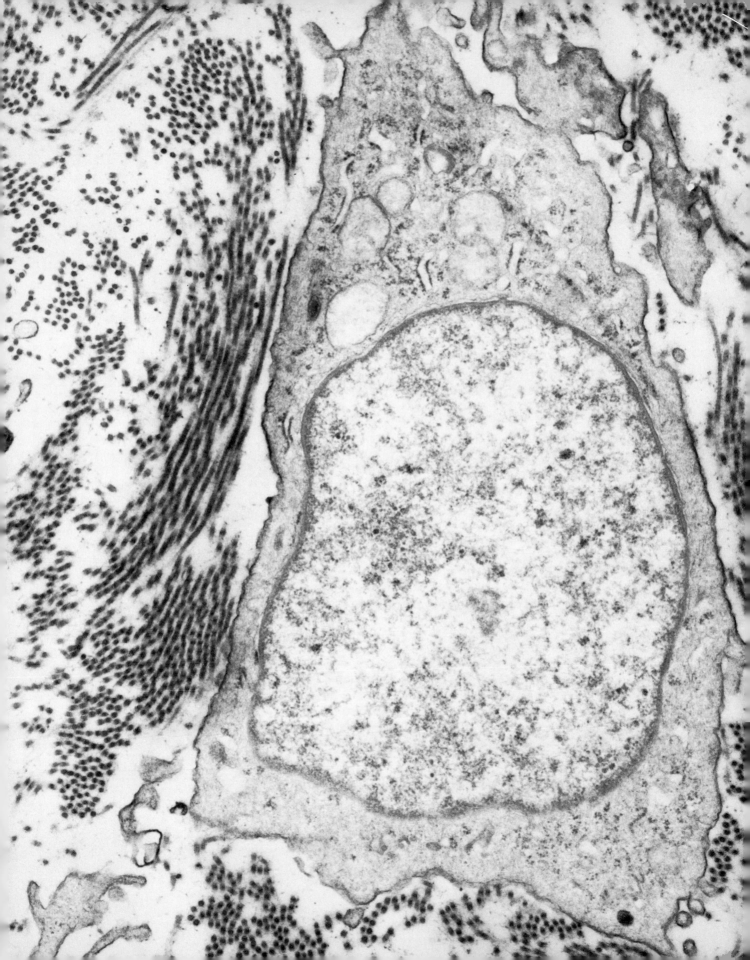

Osteosarcoma

FIGURE 156. This high magnification electron micrograph presents an intercellular zone of an osteosarcoma and a portion of cytoplasm of a malignant osteoblast. The collagen fibers are cut obliquely. A focal, characteristic deposit of hydroxyapatite is present along the collagen fibrils, obliterating its structure. (47,600×)

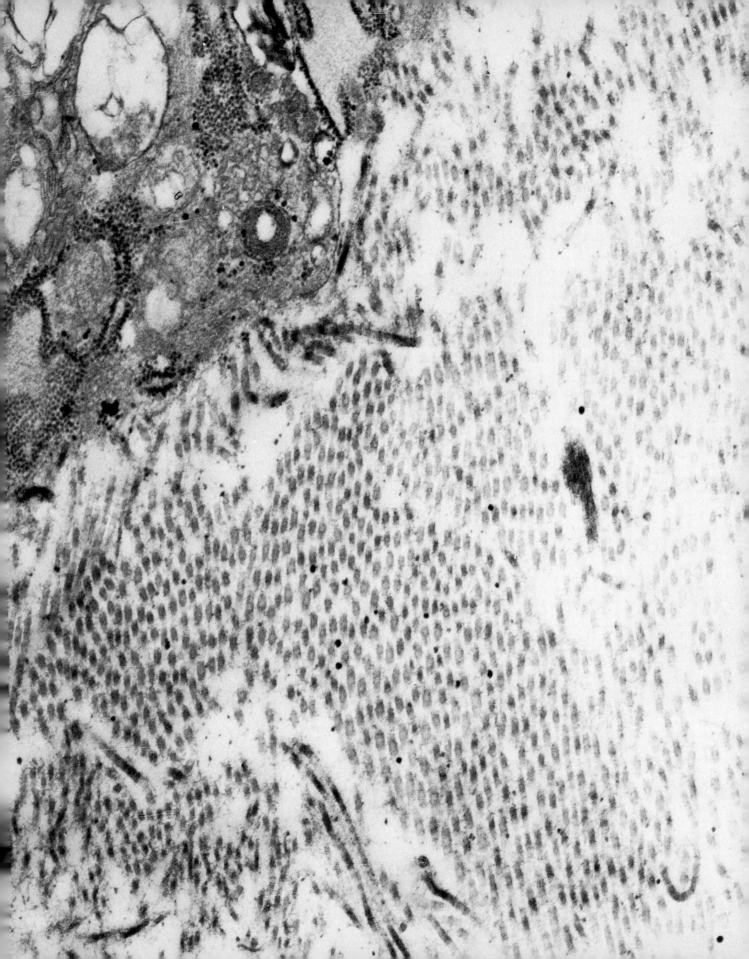

Osteosarcoma

FIGURE 157. A high magnification of the extracellular compartment shows pale and dark stained mature collagen fibrils grouped into fasciles of different width. The individual fibers vary a little in width, and most of them are oriented in the same direction. Each unit displays a uniform periodic bending pattern. Collagen fibrils predominate in dense connective tissue and are present in the extracellular space. Intracellular collagen fibrils have been found in a variety of tumors (osteosarcoma, liposarcoma, malignant fibrous histiocytoma, rhabdomyoscarcoma), not only in the fibroblastic cells but within the cytoplasm of these tumor cells as well as within the cytoplasm of primitive tumor cells. (47,300×)

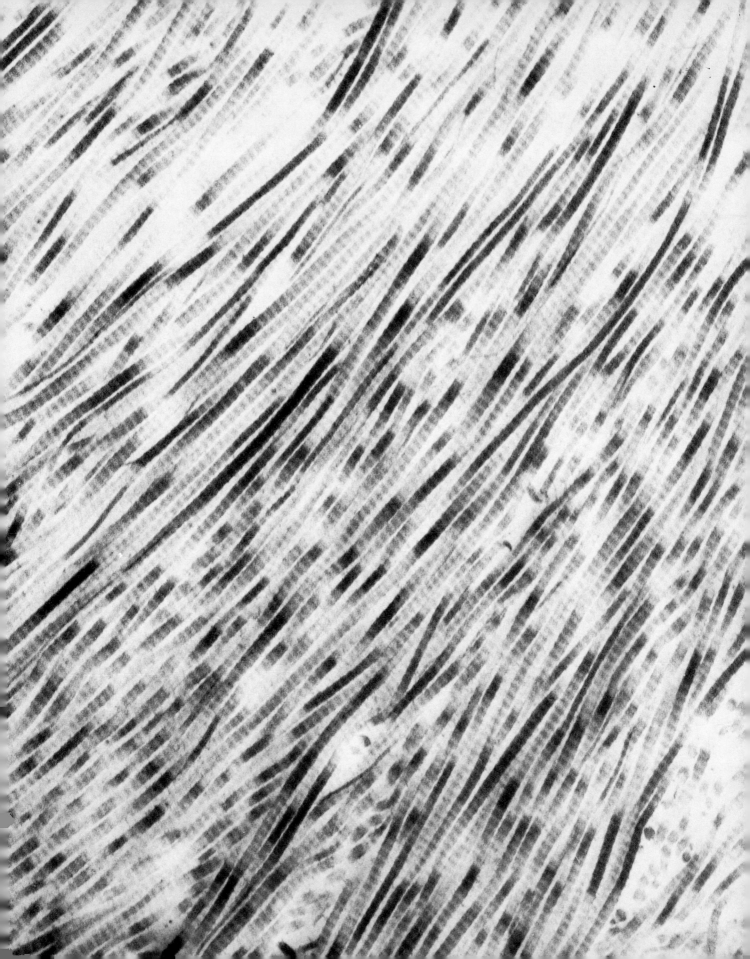

Chondrosarcoma

FIGURE 158. Several widely spaced chondrocytes occupy lacunae scattered in an abundant loose fibrillar matrix containing small dense matrix bodies. The lacunae appear as a relatively clear pericellular zone outlined by a localized compression of the fibrillar matrix. Often the lacunae contain two cells. The malignant chondrocytes are racket-shaped. The cell membranes display irregularly scalloped outlines with slender microvillous projections into the lacunae. The nucleus is eccentric. The most characteristic feature is the presence of a well-developed RER occupying most of the cytoplasm. The RER appears as a lamellar array or as vesicles whose cisternae contain a finely granular secretory product. Glycogen forms large aggregates or lakes located predominantly at the periphery of the cytoplasm. Few interspersed mitochondria are observed. (9500×)

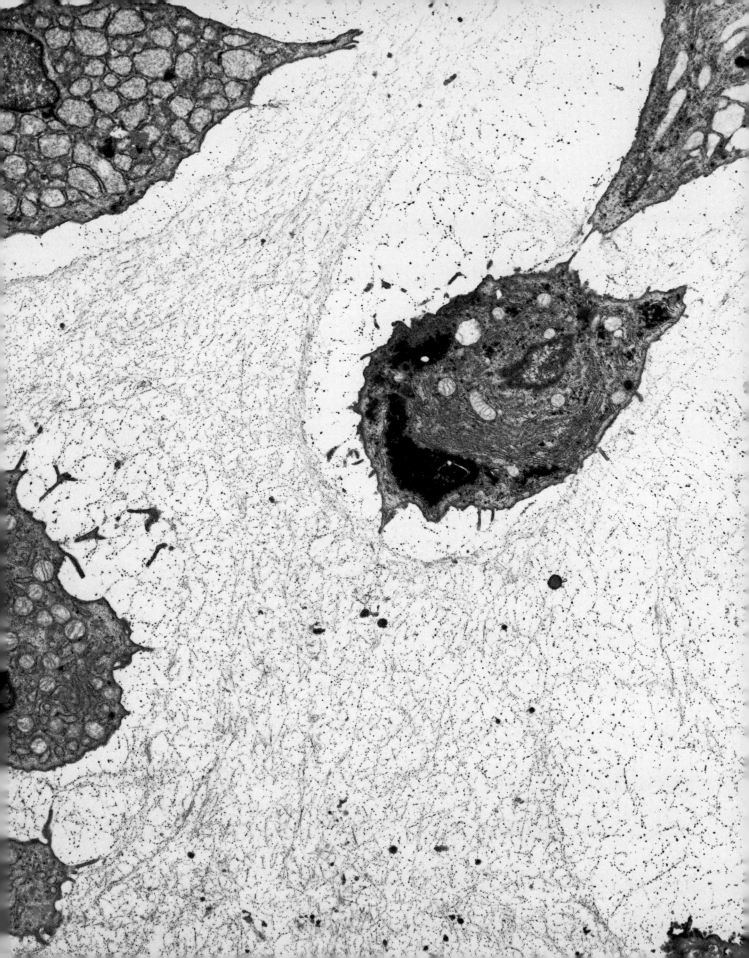

Chondrosarcoma

FIGURE 159. This cell in mitosis is from a chondrosarcoma. The predominant feature is glycogen aggregates. The lacunae as well as the fibrillar and dense matrix bodies are clearly seen. (14,500×)

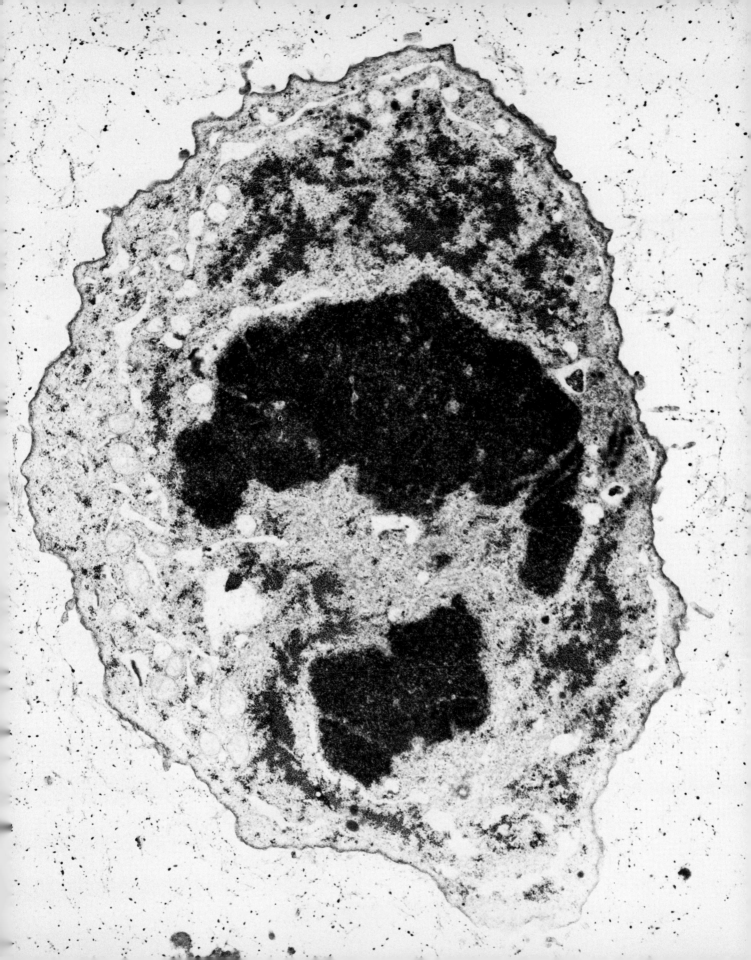

Chondrosarcoma

FIGURE 160. This micrograph is a high magnification of a portion of a malignant chondrocyte. The lacuna is clearly seen as an obvious pericellular zone containing few fibrillar, dense matrix bodies into which the cytoplasmic villi project. The cytoplasm is rich in RER which appears as arborizing lamellae in the perinuclear cytoplasm and as irregular large spaces filled with floccular substance in the outer portion of cytoplasm. More peripherally, large accumulation of glycogen and glycogen lakes are seen. The nucleus is eccentricly indented and contains a coarse chromatin substance (heterochromatin) and two nucleoli. (15,200×)

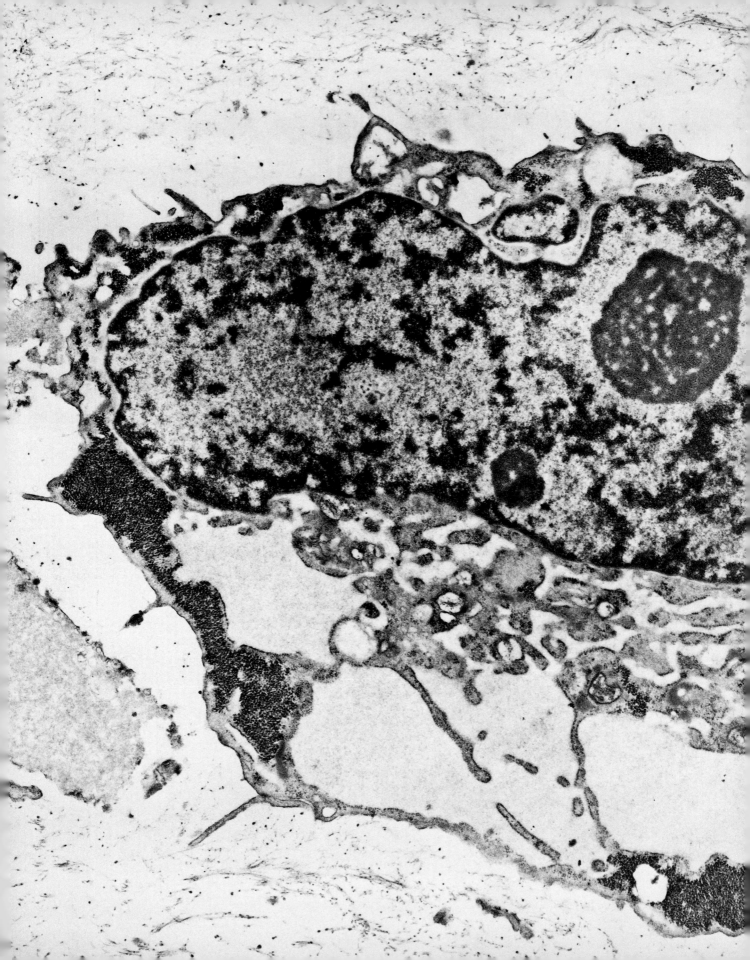

Ewing's Sarcoma

FIGURE 161. The cells in Ewing's sarcoma appear to be round with cytoplasm rich in glycogen. The cytoplasm also contains mitochondria and both smooth and rough endoplasmic reticulum. The nucleus shows nucleoli and prominent clumped chromatin which is partly arranged around the nuclear membrane. (17,500×)

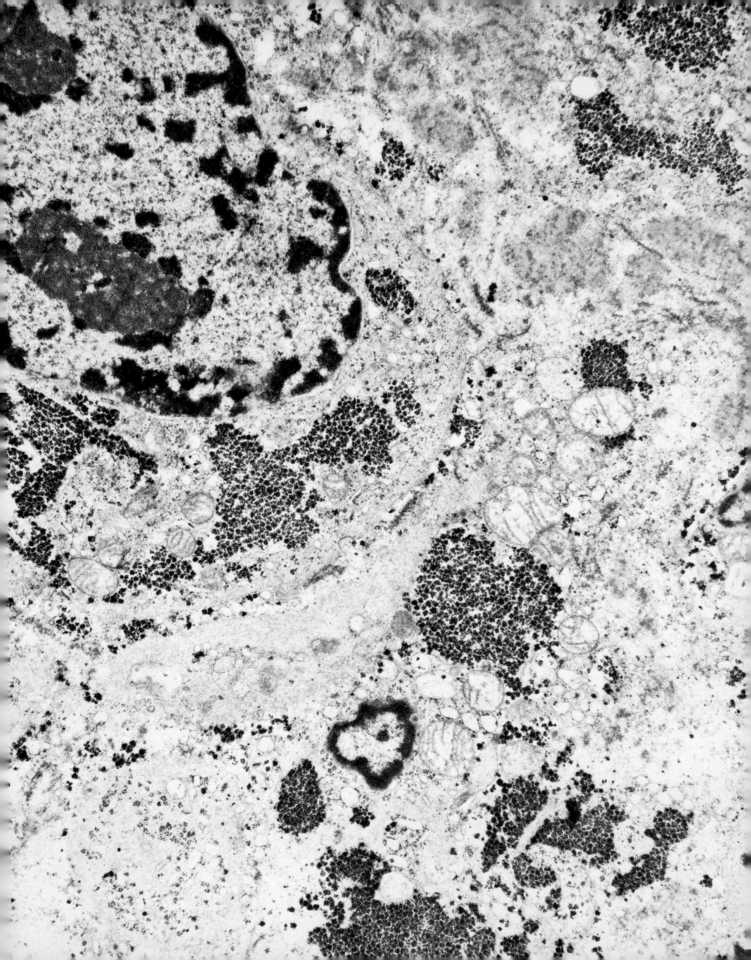

Synovial Sarcoma

FIGURE 162. Two types of cells are shown here. There is the large epithelial cell on the top with cytoplasm rich in mitochondria. The lower middle cell is more fibroblastic with cytoplasm rich in endoplasmic reticulum. Occasional desmosomes are seen in the membrane of the lowermost cell. (27,900×)

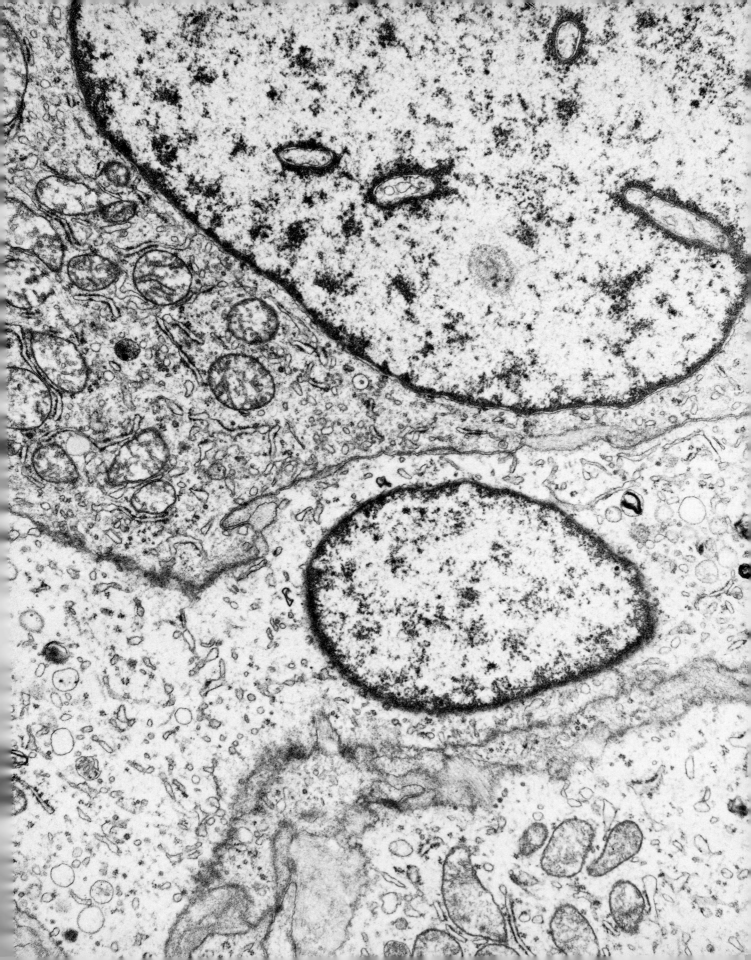

Higher Magnification of Synovial Sarcoma

FIGURE 163. This is a close-up of epithelial cell cytoplasm which is rich in dilated mitochondria. It also has some smooth surfaced and rough surfaced endoplasmic reticulum. (66,000×)

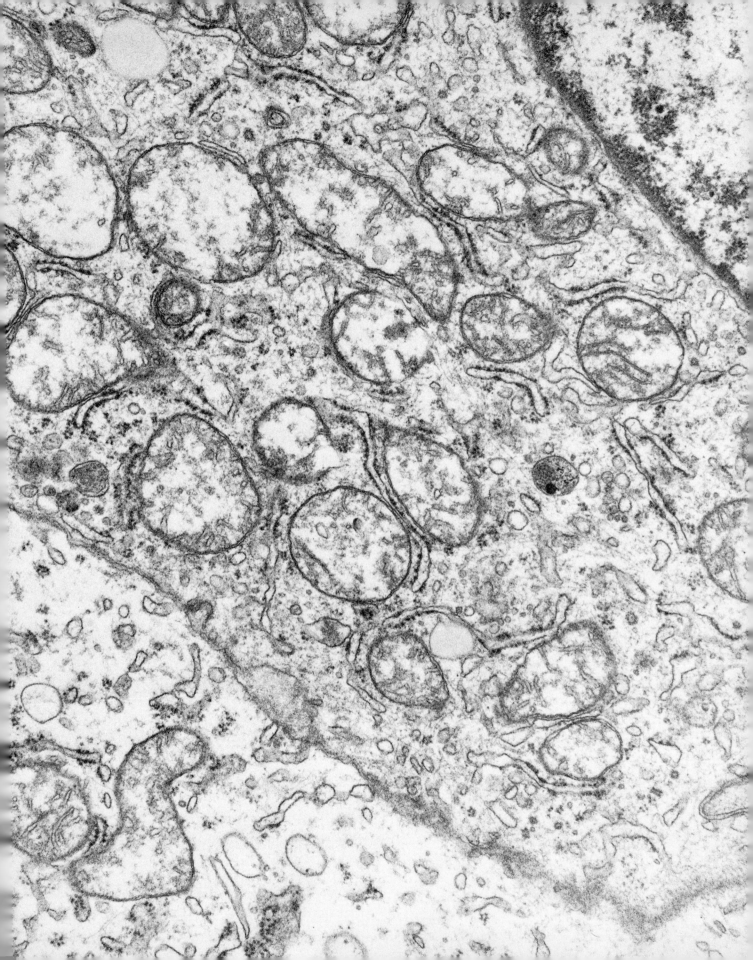

**Metastatic Squamous Cell Carcinoma
In the Femoral Head**

FIGURE 164. The tumor cells have closely apposed cell borders which frequently show specialized intercellular junctional complexes, the desmosomes. The cytoplasm discloses bundles of filaments, the tonofilaments, which often terminate in the desmosomes. (24,000×)

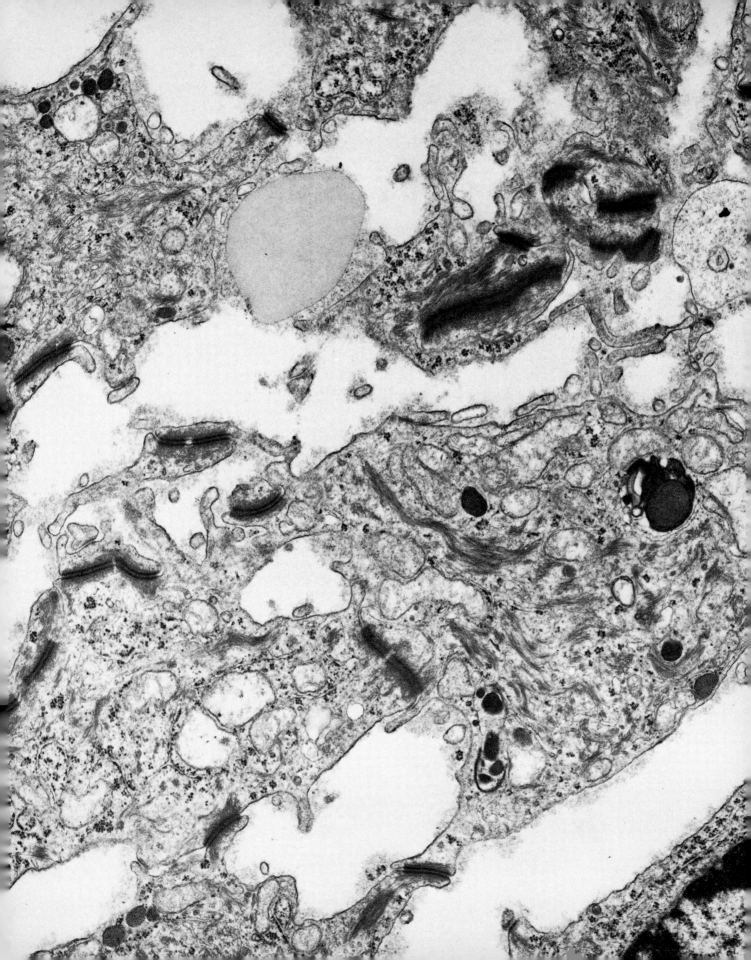

20 Reticuloendothelial System

Schaumann Body

FIGURE 165. This is a portion of an epithelial cell in noncaseating granuloma in a lymph node from a case of likely sarcoidosis. The impressive laminated body is the ultrastructural correlate of the Schaumann body that seen by light microscopy. The body is a lysosomal residual body. Primary lysosomes can also be found in these epithelial cells which are thought to be derived from blood monocytes. (25,500×)

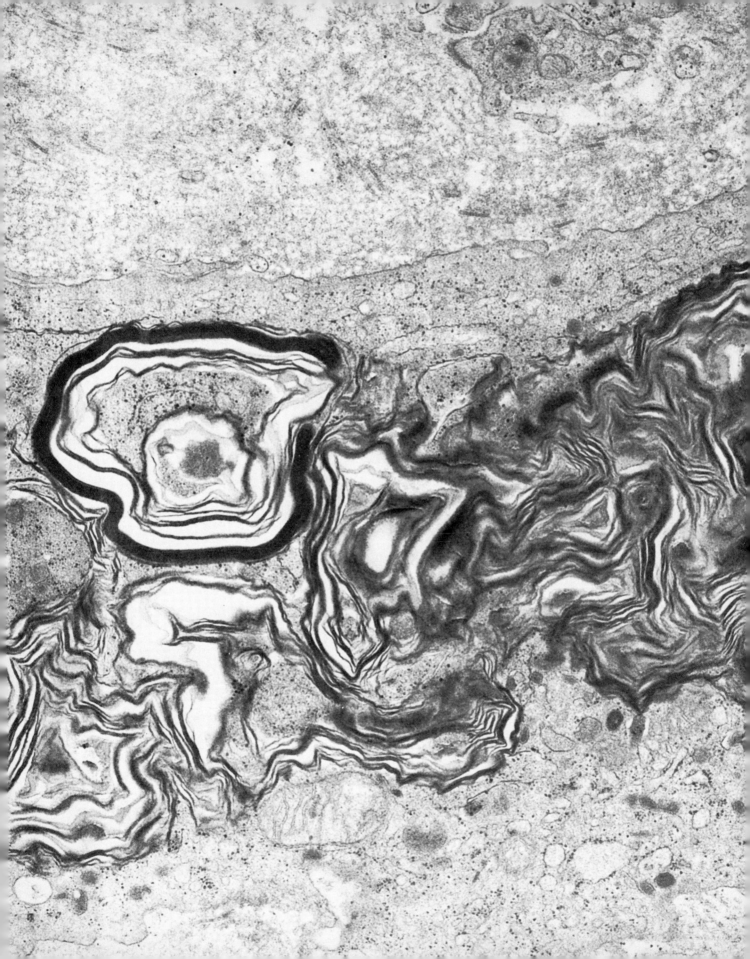

**Poorly Differentiated Lymphocytic
Lymphoma—Lymphoblasts**

FIGURE 166. The cells have enlarged, slightly indented nuclei with prominent nucleoli. The latter are often surrounded by nucleoli associated chromatin. The cytoplasm is scant in organelles, has few mitochondria and RER, and has many free ribosomes and polyribosomes. (12,800×)

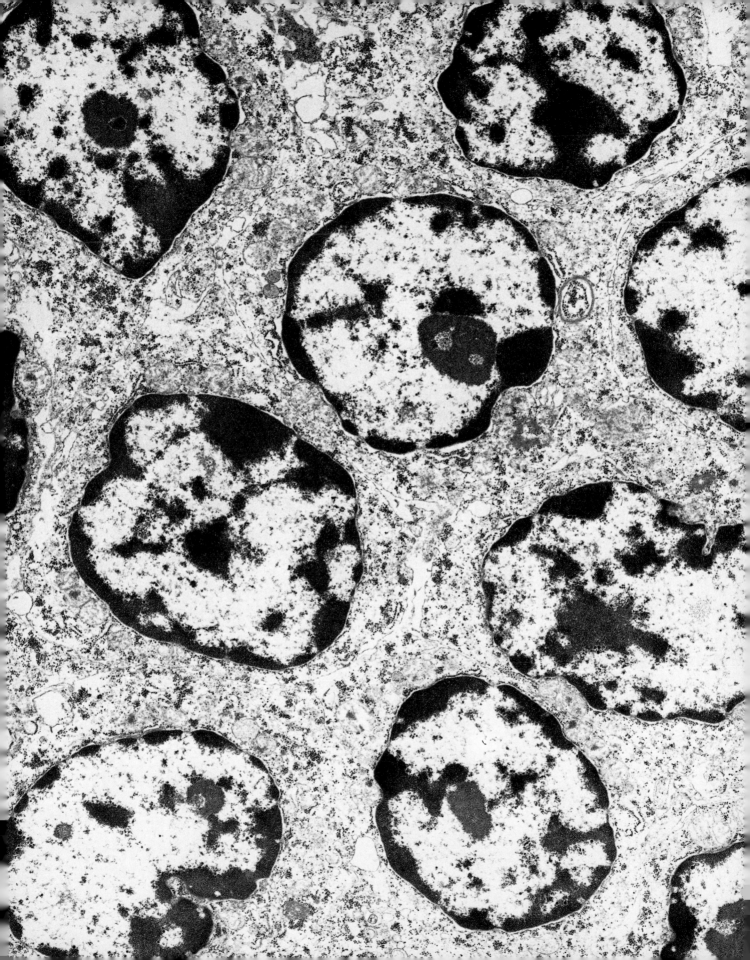

Reed-Sternberg Cell

FIGURE 167. The cell in this electron micrograph shows the essential ultrastructural pictures of the Reed-Sternberg cell which is most often associated with Hodgkin's disease. The nucleus appears bilobed with prominent nucleoli and with an internuclear bridge. The presence of these nuclear features is very dependent on the plane of sectioning through a particular cell. The cytoplasm is distinctive for its few organelles but numerous free ribosomes. ($25,000\times$)

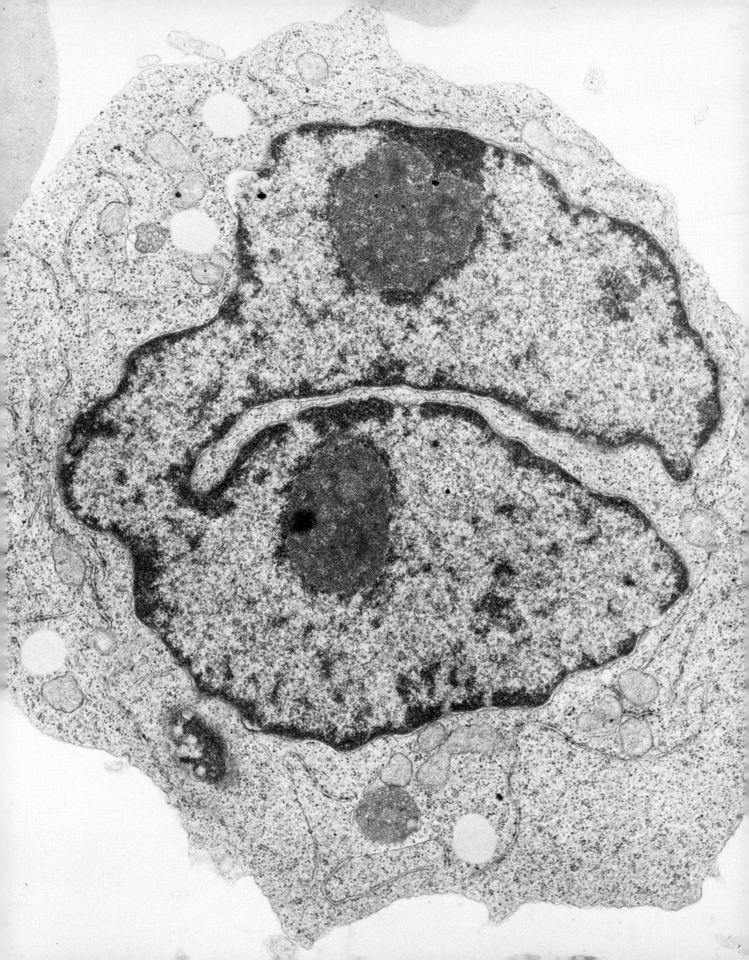

Extramedullary Hematopoiesis in the Spleen

FIGURE 168. There are usually three cellular components, namely, erythroblasts, megakaryocytes, and immature granulocytes, in extramedullary hematopoiesis in the spleen. The following two illustrations are from a case of chronic lymphocytic leukemia with thrombocytopenia.

This picture shows three polychromatophilic erythroblasts having a few mitochondria and diffusely scattered polyribosomes. Red cell maturation is manifested by a progressive increase in peripheral chromatin, together with a decrease in the cytoplasmic content of ribosomes and mitochondria.

A lymphocyte (lower left) shows anastomotic endoplasmic reticulum which appears continuous with the outer nuclear envelope. Fibrils within the endoplasmic reticulum, occasionally extending into the perinuclear space, are present. (16,400×)

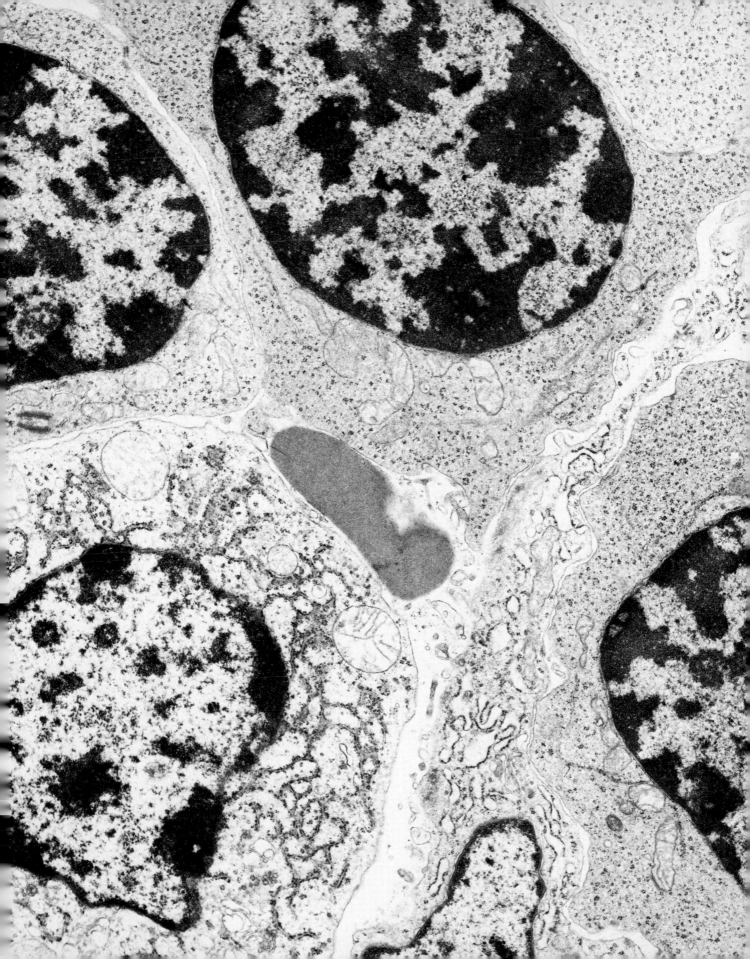

Extramedullary Hematopoiesis in the Spleen

FIGURE 169. A part of a megakaryocyte shows characteristic findings of a demarcation membrane system and electron-dense membrane-bound granules (azurophilic). Demarcation membranes appear to be continuous with the surface membrane of the megakaryocyte. (27,500×)

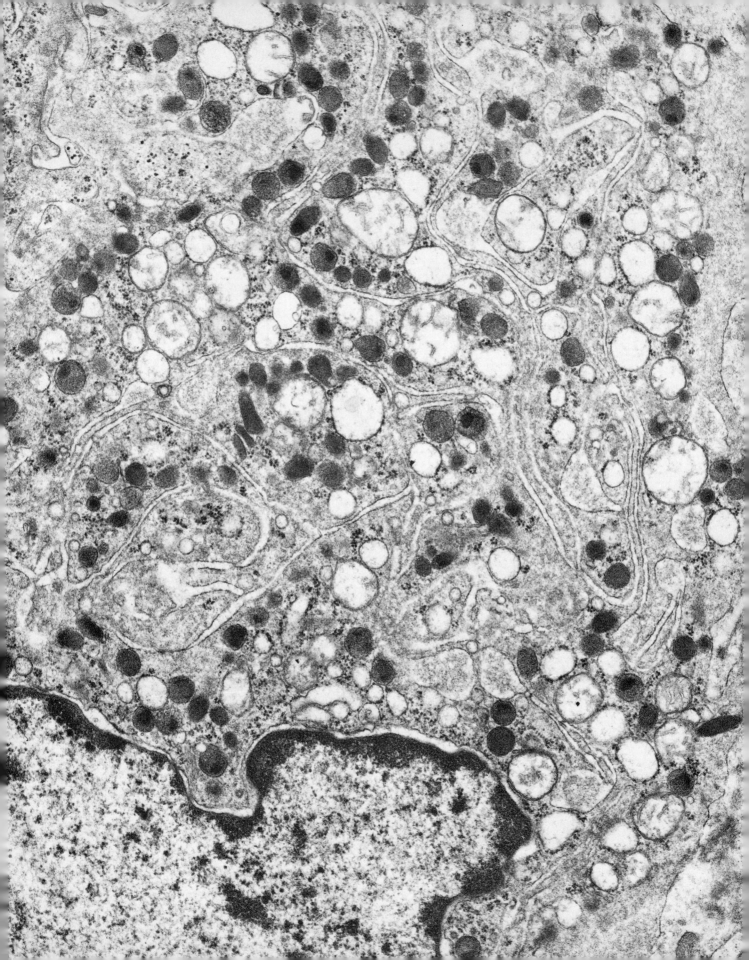

Sickled Red Blood Cells

FIGURE 170. Everyone of the numerous red blood cells throughout this micrograph contains fine electron-dense filaments. They are precipitated as hemoglobin S, abnormal hemoglobin which is hereditarily present in this patient. The precipitation of the hemoglobin in this biopsy specimen was initiated by the deoxygenation during biopsy and glutaraldehyde fixation. Deoxygenation in vivo also causes precipitation and filament formation that leads to the sickling deformation of the red blood cells and the consequent pathological manifestations of the disease. A higher magnification of the filaments is contained in the next figure. (10,000×)

340

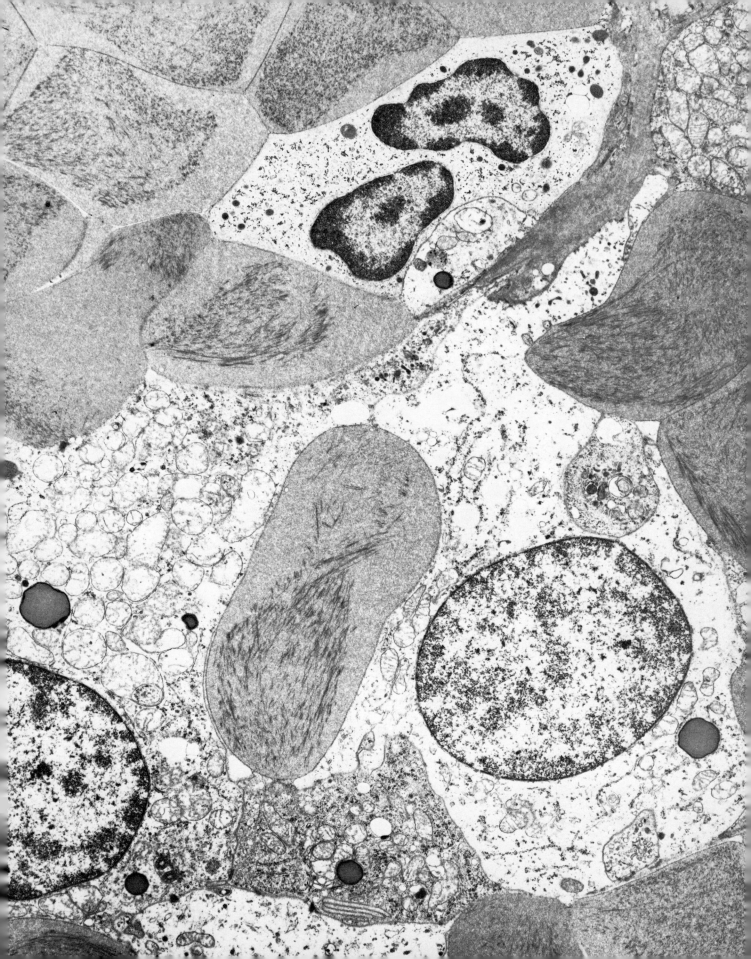

Hemoglobin S Filaments in a Sickled Red Blood Cell

FIGURE 171.   This is a high magnification of one of the red blood cells of the preceding figure. The fine individual filaments of precipitated hemoglobin S are easily noted. (49,000×)

Auer Rod in Acute Granulocytic Leukemia

FIGURE 172. A leukemic cell contains a characteristic Auer rod. The Auer body is produced by fusion of large azurophilic granules which are normal features of granulocytic blood cells. The Auer body can be rod or needle-shaped, and stains azurophilic. They generally measure 0.2 to 0.7 μm in length. Auer bodies give positive oxidase, peroxidase, PAS, RNA, and acid phosphatase reactions. They are most frequently found in acute granulocytic leukemia. (29,900$\times$)

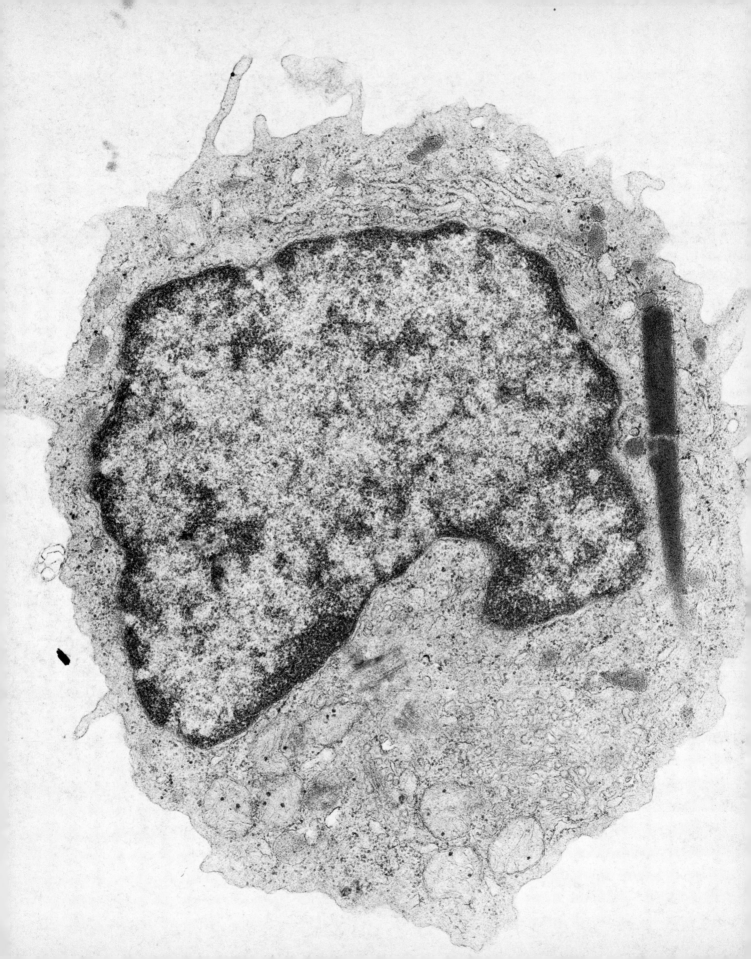

Higher Magnification of an Auer Rod

FIGURE 173. This high magnification of the preceding photograph shows a characteristic Auer rod composed of a linear periodic substructure. (88,400×)

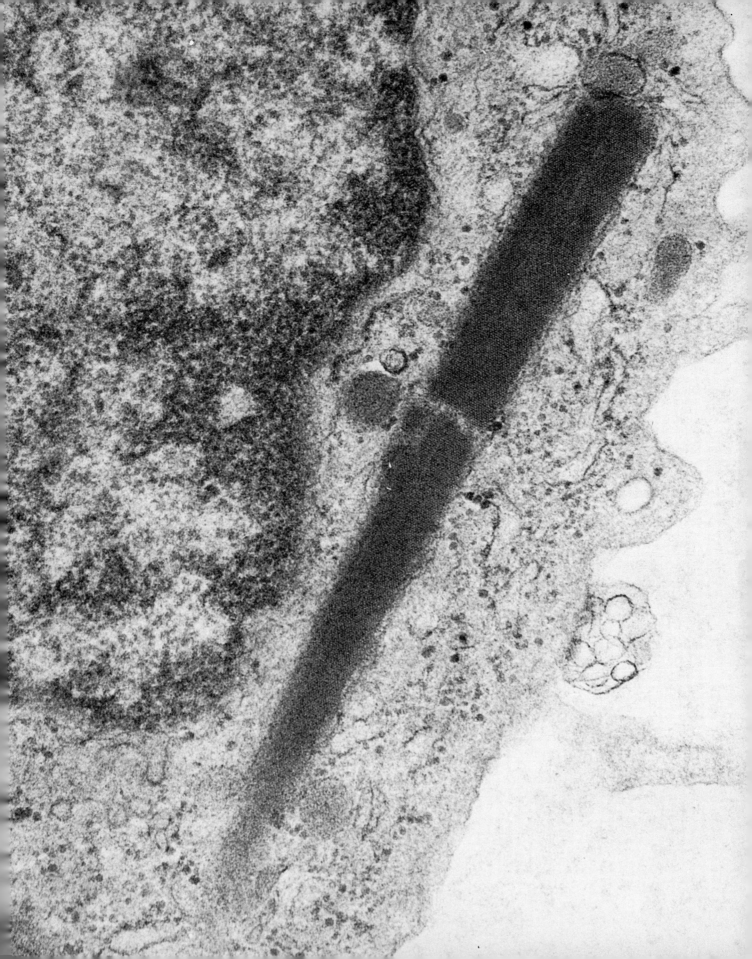

Hairy Cell in Hairy Cell Leukemia

FIGURE 174. The cell shows large irregular villous cytoplasmic projections with the nucleus having condensed peripheral nuclear chromatin and a distinct nuclear membrane. The cytoplasmic organelles consist of mitochondria, rough endoplasmic reticulum, and ribosome–lamellae complexes. (34,700×)

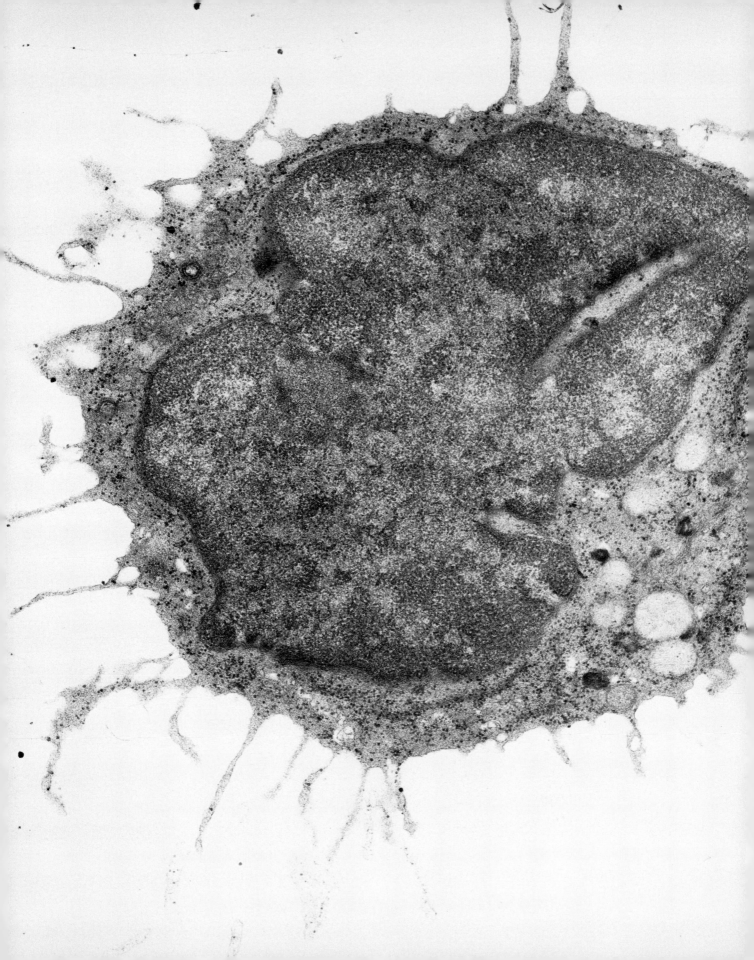

Ferruginous Mitochondria in Erythroblasts from Bone Marrow in Sideroblastic Anemia

FIGURE 175. The nucleated cell in the center of the micrograph is an immature red blood cell called an erythroblast. It contains abnormal mitochondria having very electron-dense iron which is located in the inner matrix but not in the intracristal space nor between the outer and inner membrane. The dense material free in the cytoplasm is glycogen which is also an abnormal finding in an erythroblast.

The less dense cytoplasmic particles are free ribosomes. Mature red blood cells are seen at the periphery of the electron micrograph. (39,765×)

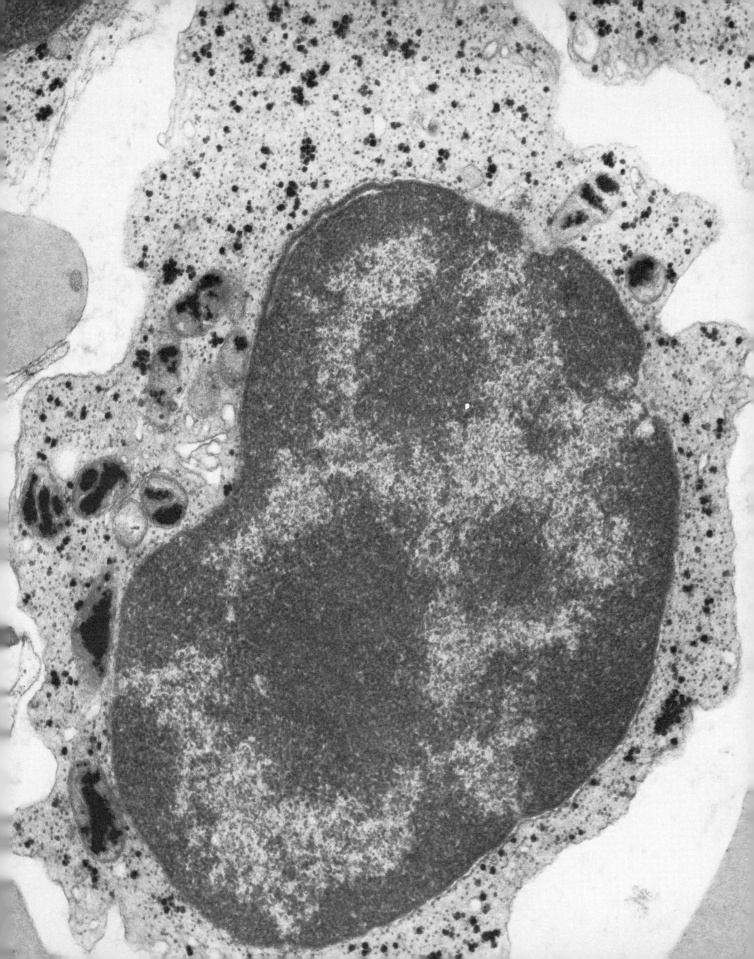

Iron

FIGURE 176.   Iron, usually in a protein complex known as ferritin, characteristically presents itself ultrastructurally as specific 10 nm particles. Because the iron content of the particles makes them highly electron-dense, the particles are more easily distinguished from contamination and other artifacts in unstained rather than stained sections. The iron in this unstained material is easily recognized and can be distinguished in two separate compartments. One compartment is the cytoplasmic matrix where the iron particles are dispersed in an electron lucent background; the other compartment is the membrane-bound, concentrated accumulation of iron particles defined as a siderosome and located at the right edge of the figure. Iron particles are both a normal and abnormal ultrastructural feature depending on the cell. (110,000×, unstained except for osmium tetroxide fixation)

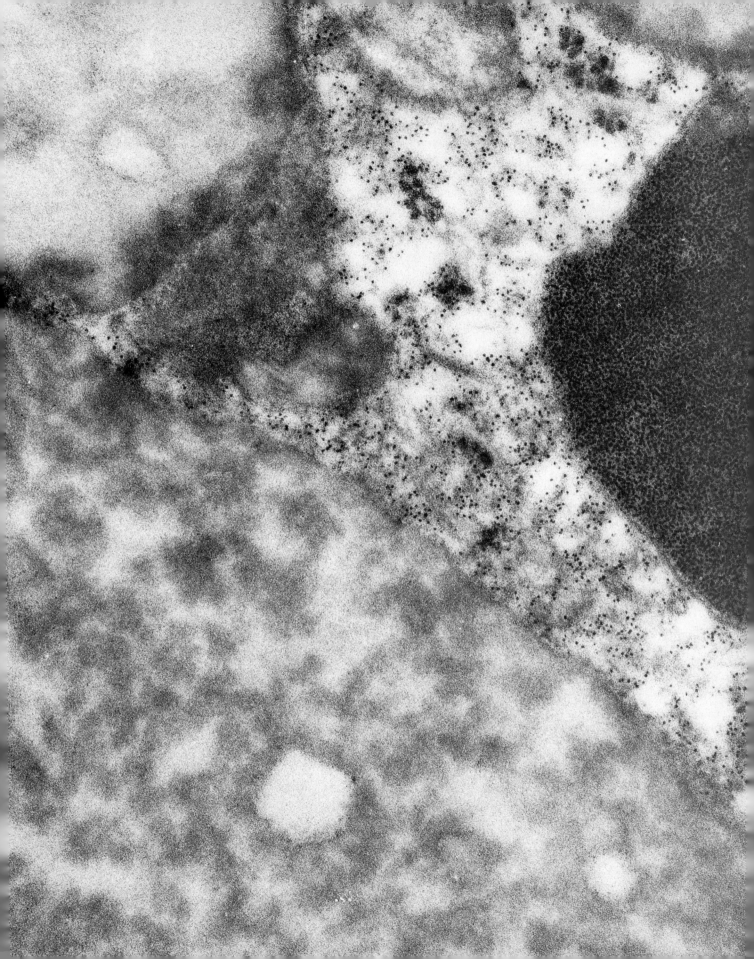

Eosinophil

FIGURE 177. The eosinophils in this electron micrograph are characterized by their dense lysosomal granules which contain a central dense crystalloid. The crystalloid is different in different species. The crystalloid in human eosinophils is noted for its variability. The eosinophil is a normal but infrequent component of blood and tissue. An inflammatory reaction that includes many eosinophils is suggestive of an immunologic or allergic reaction. (14,500×)

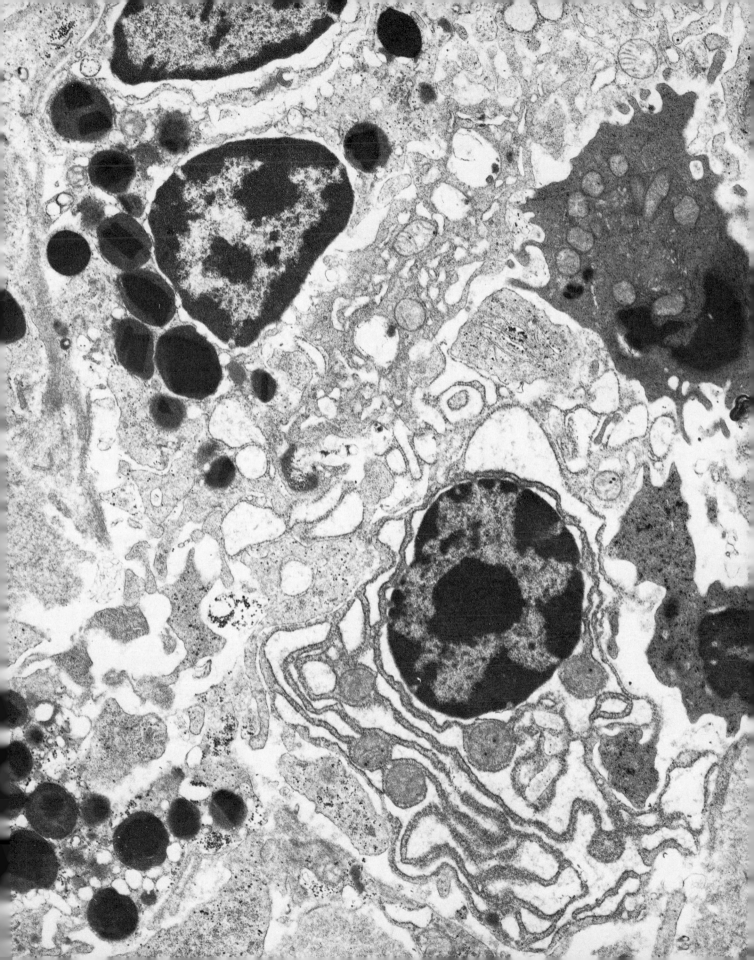

Plasma Cell

FIGURE 178. The plasma cell is characterized ultrastructurally by a cytoplasm of predominantly rough endoplasmic reticulum and by a polarized nucleus with periodic clumps of heterochromation along the nuclear membrane. The lumen of the endoplasmic reticulum is filled with an amorphous material which is presumably the immunoglobulin that is synthesized and secreted by the plasma cell. The occurrence of plasma cells in an inflammatory response is associated with a chronic inflammation and, of course, an immune response. (24,800×)

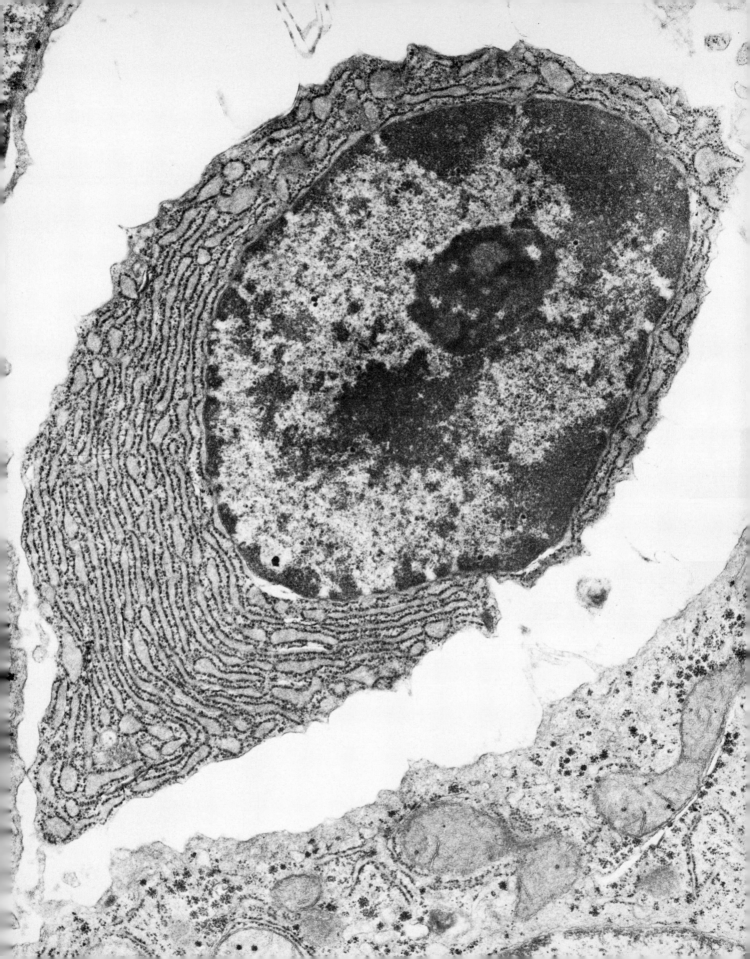

Russell Body in a Plasmacytoid Cell

FIGURE 179. A plasmacytoid lymphocyte in a lymph node biopsy specimen from a two-year-old boy with systemic lupus erythematosus and generalized lymphadenopathy is shown. Note the several large crystalline-like structures (arrow) among the rough endoplasmic reticulum. These structures are more frequently seen in plasma cells, particularly in plasma cell dyscrasia, but they can be seen in lymphoproliferative disorders like this case. (29,400×)

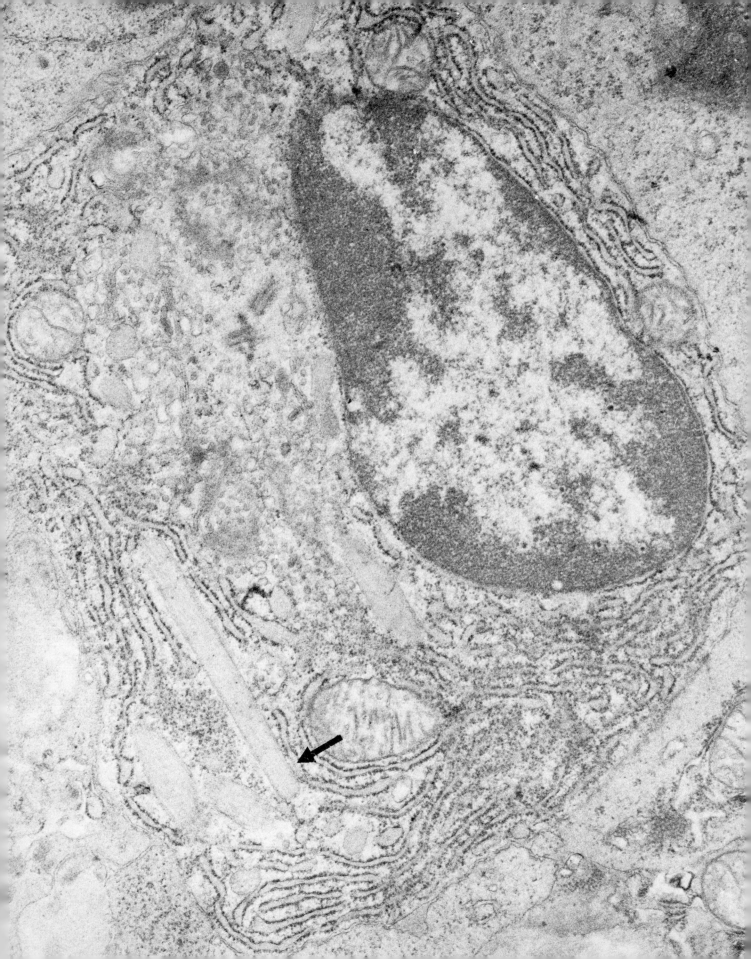

Polymorphonuclear Leukocyte Phagocytizing Bacteria

FIGURE 180. The various stages in the initial phagocytosis of bacteria is exemplified in this polymorphonuclear leukocyte. The stages range from simple attachment of a bacterium to the cell surface to complete incorporation of the intact bacterium into an intracytoplasmic vacuole. The lysosmal digestion of the bacterium is not seen in this early stage. (16,400×)

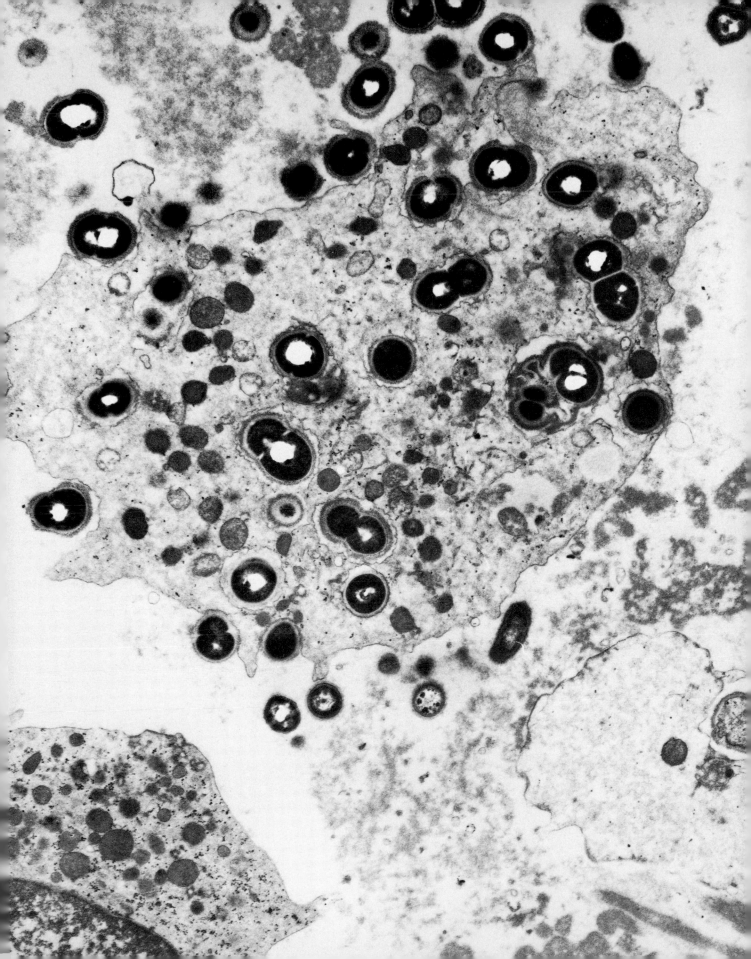

21 Skin

Molluscum Contagiosum

FIGURE 181. The viral particles are present in the cytoplasm of epidermal cells. The nucleus is displaced by distended cytoplasm. (3600×)

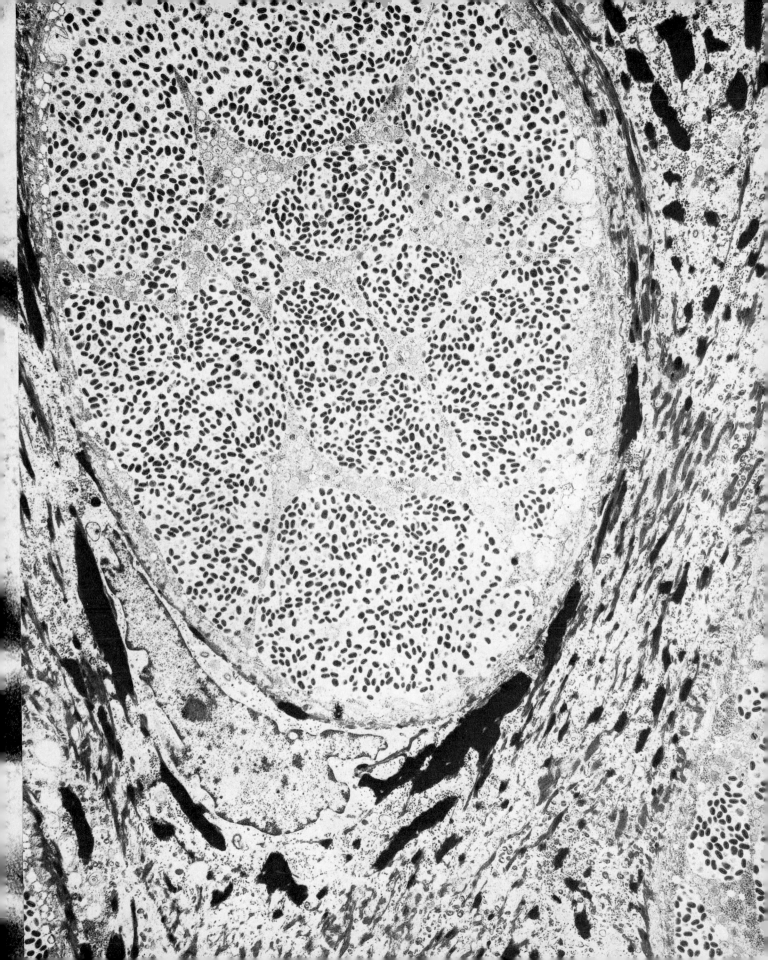

Herpes Virus Infection of the Skin

FIGURE 183. A keratinocyte contains numerous virus particles in the nucleus and in the cytoplasm. Many virus particles have in their center a small, electron-dense core or nucleoid that is surrounded by the capsid. Others show only an empty capsid. Most virus particles in the cytoplasm are enveloped by another outer coat which is derived from the nuclear membrane. Ultrastructurally, the virus of *Herpes hominis* is indistinguishable from that of *Herpes varicellae*. (46,700×)

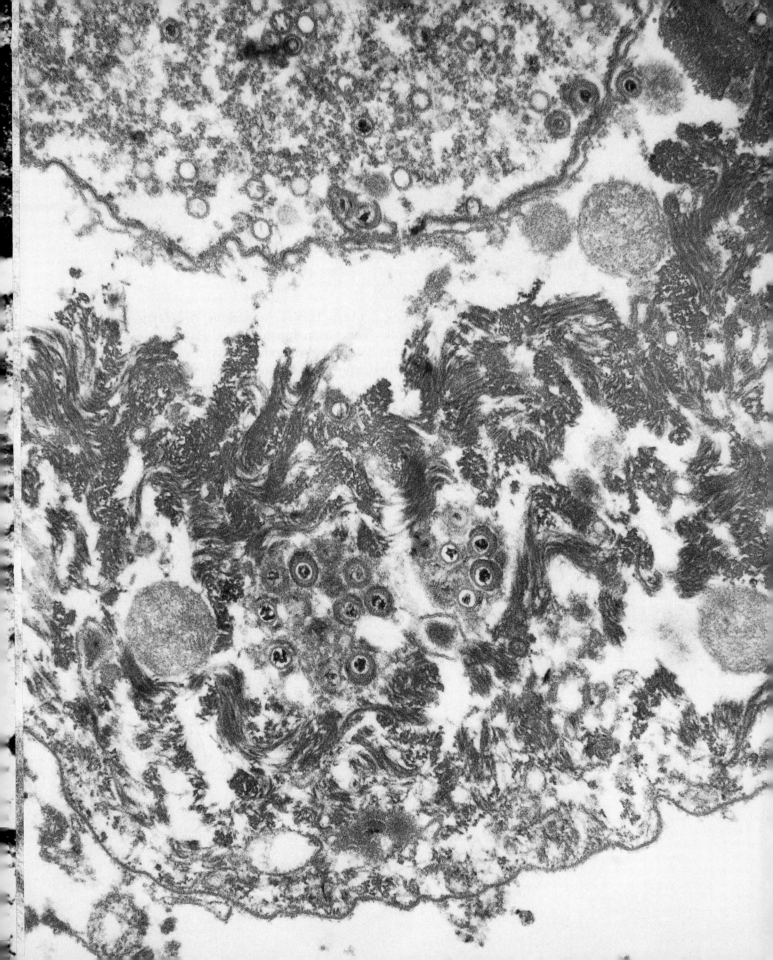

Higher Magnification of *Treponema pallidum*

FIGURE 186. The cytoplasm of the spirochetes contains ribonucleoprotein particles and tubules and is delineated by a bilaminar cell membrane, which is not preserved here because of Formalin fixation and paraffin embedding. (81,800×)

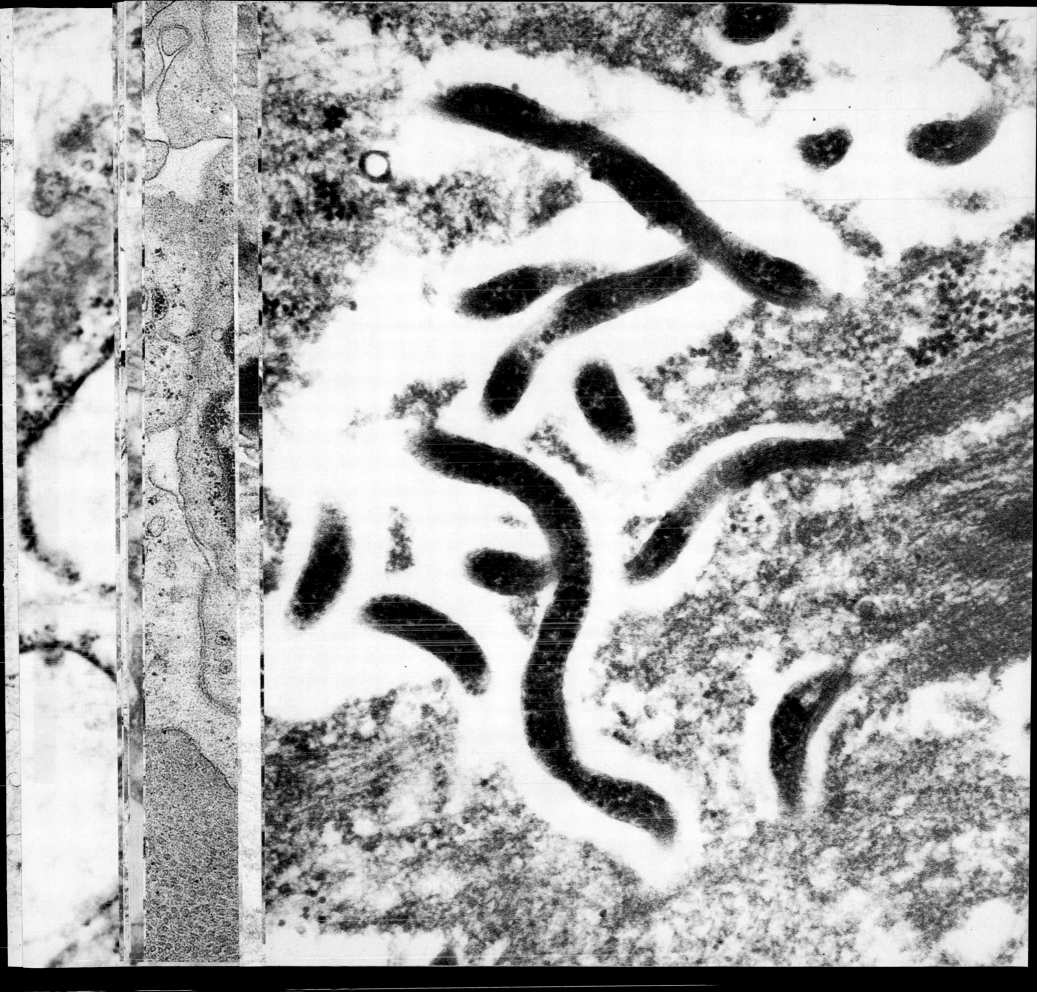

Neurofibrosarcoma

FIGURE 200.　The neoplasm arising from the popliteal nerve was removed from a youngster with neurofibromatosis. The neoplastic cells have a relatively large ovoid or elongated nucleus with an occasional small peripheral nucleolus. The chromatin pattern is finely granular and without obvious clumping. In the cytoplasm there is a modest amount of rough endoplasmic reticulum, ribosomes, and mitochondria. The rough endoplasmic reticulum is frequently distended with proteinacious material. A small Golgi apparatus is occasionally seen near the nucleus. Microfilaments are rarely seen. The individual cells are surrounded by considerable amount of collagenous fibrous tissue. Note the periodicity of the collagen fibers. Histologically this neoplasm is extremely similar to fibrosarcoma. However, its association with the deep seated large peripheral nerve and a history of neurofibromatosis is helpful in diagnosis. (11,100×)

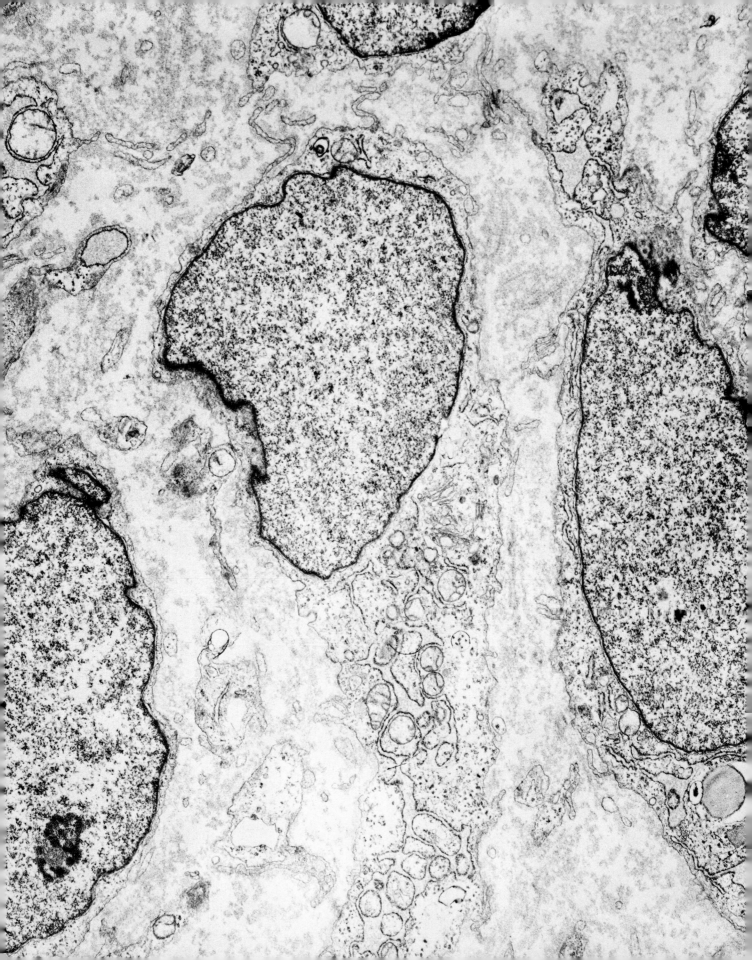